A Manual of Practical Entomology
(Field and Laboratory guide)

(3rd Revised and Enlarged Edition)

M.M. Trigunayat

Ph.D., FESI, FZSI

Head, Department of Zoology

Govt. R.D. Girls (PG) College,

Bharatpur - 321001 (India)

SCIENTIFIC® PUBLISHERS

New Delhi • Jodhpur

Published by:
Scientific Publishers

5 A, New Pali Road, P.O. Box 91
Jodhpur - 342 001 Rajasthan (India)

4806/24, Ansari Road, Daryaganj,
New Delhi - 110 002 (India)

E-mail: info@scientificpub.com
Website: www.scientificpubonline.com

Frist Edition : 2002
Second Edition: 2009

ISBN: 978-81-72339-82-1 (H/B)
eISBN: 978-93-86102-02-7 (E/B)

Printed in India

PREFACE (3rd Edition)

I have received commendable response from the readers from all corners of country and abroad. The book was first published in 2002 and since then the trend of the readers has been increasing. The efforts have been made to make the book more comprehensive and several other exercises have been included. In the present edition general account on pests like Red pumpkin beetle, Thrips and scale insects have been incorporated. I would feel pleasure if the readers can communicate the mistakes and other limitations left over.

Dr. M.M. Trigunayat
E-mail: mm_trigunayat@yahoo.com

PREFACE

The present book is an outcome of the long need felt by the students of Entomology. Though there is no dearth of literature on such a popular subject like entomology but hardly any book is available as laboratory guide which can display all exercises mentioned in their syllabi in comprehensive and concised manner, if at all available, they hardly envisage the whole aspects of practical syllabi. Thus, this manual is unique in this respect as it envisage almost all broad headings of the practical syllabi of M.Sc. classes viz. Collection and preservation, Conspectus of insect orders, Identification of insects with the help of keys, Applied entomology, Insect physiology, ecology, toxicology and behavioural exercises, Histological slides, Dissections and Microtomy. A note on field report is also being given, as to how a student should prepare a report of the area visited by him. Originality of the presentation and style is claimed.

A number of references and books have been consulted in depth during the preparation of manuscript. Though it is not possible to enlist them but a few are being listed in the `References'. I have received the help of different personales in various ways (encouragement, suggestions and blessings) who either have been my colleagues, learned teachers, research personales at different part of the country, friends and students. I greatly acknowledge the help rendered by Prof. N.P. Singh (Jaipur), Prof. Irfan Ahmad (Aligarh), Prof. H.S. Rose (Jalandhar), Prof. B.N. Pandey (Bodhgaya), Prof. S.K.A. Rizvi (Aligarh), Dr. Farida (Jaipur), Mr. Raj Mohan Meena, Mr. Vivek Singh Maharia (Jhunjhunu), Prof. K.K. Sharma (Ajmer), Prof. Girish Sharma, Agriculture Research Station, Durgapura, Jaipur, Late Prof. R.K. Gautam, (Agra), Prof. J.S. Kirti (Patiala), Prof. S.C. Dwivedi (Jaipur), Dr. Rajender Kaushik (Jaipur), Prof. Jaimala Sharma (Jaipur), Dr. Kavita Sahni (Jaipur), Dr. Arti Mathur (Udaipur), Prof. O.P. Agarwal (Gwalior), Dr. Rajesh Sharma (Alwar), Dr. Ashok Purohit (Jodhpur), Dr. V.B. Awasthi (Jodhpur), Dr. N.S. Rathore (ZSI, Jodhpur), Dr. S. Shanmugavel (Tuticorin), Dr. Bharti

Prakash, Dr. Kalpana Sharma (Ajmer), Dr. Neeta Jain (Kota), Dr. Uma Shankar Sharma (Udaipur) My colleagues at Bharatpur have been helpful indirectly in one way or other. I am indebted to my wife Neerja and children Kritika and Mrigank for their endless patience shown during the preparation of this manuscript. I am thankful to Shri Pawan Kumar Sharma and Mr. Tanay Sharma, Scientific Publishers (India), Jodhpur for bringing out this book. Last but not the least, thanks are due to my niece Tanu and my daughter Kritika for drawing some of the figures.

Suggestions of the readers, are invited.

Dr. M.M. Trigunayat

CONTENTS

CHAPTER - 1

INSECT COLLECTION AND PRESERVATION

INSECT COLLECTION

Gaining mastery on the technique of collecting and preserving insects for scientific study is a must for entomology student. One can be acquainted with the different groups of insects by individual collections. With little sincerity and efforts excellent collections can be made. The specimens can be identified upto order, family and species level. It is not difficult to identify the insects upto family but upto the species level we require special books or they can be sent to specialists and certain institutions in different parts of the world.

Insects can be collected from their natural habitat at any season of the year. We require certain appliances for collecting insects. The first hand information of these equipments is given below.

1. INSECT NETS

Nets are supplied by the dealers of Scientific instruments or Entomological equipments. Select a net of your own requirements.

A net consist of a cloth bag or nylon net bag, a handle and a metal ring which holds the mouth of the open bag. A ring of 38- 40 cm diameter made up of 3-4 cm thick wire is used. The ends of the ring are fitted into a groove at the end of handle. This detachable ring can allow the change of the net if it is dirty or torn out.

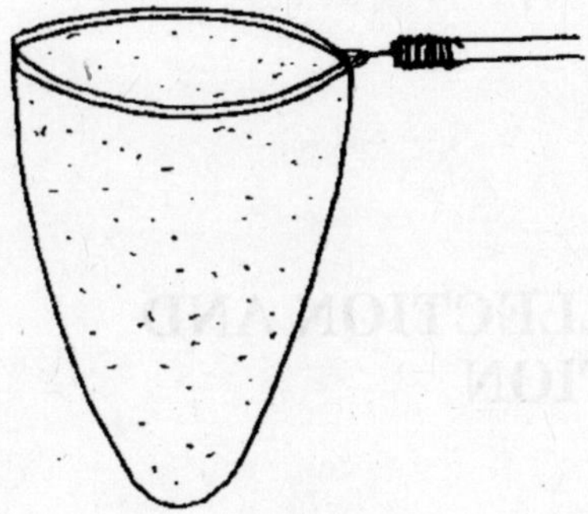

Fig. 1.1. Insect net

The depth of the bag is 75-76 cm or twice the diameter of the ring. The handle should be light but sturdy.

You can also prepare your own net on a frame of badminton racket. It is very easy and handy to sweep.

2. ASPIRATOR (POOTER)

It is made up of a specimen tube having two holes at the cork stopper, through one of which is inserted a bent glass tube and through other, a rubber tubing fitted at the end of a straight glass tube (the suction tube). The smaller insects can be sucked in by putting the inlet tube near the insects and sucking air through the rubber tube. Stored grain pests (beetles) are easily handled and collected in no time.

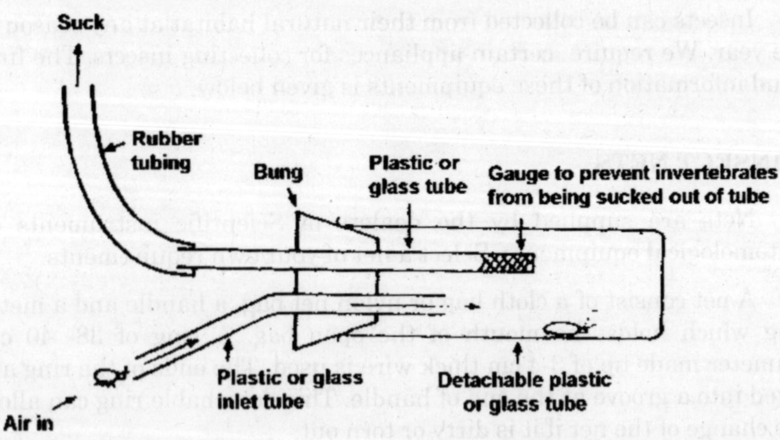

Fig. 1.2. Aspirator

3. KILLING BOTTLES OR JARS

Cyanide bottles were used for this purpose earlier but now-a-days their use has been discouraged because extraordinary care is needed to handle them. It is always safe to prepare a killing bottle of your own requirements either with ethyl acetate or benzene, chloroform, ether or carbon tetra chloride.

Take a bakelite top wide mouth jar and spread a layer of plaster of paris. Allow it to dry and saturate it with ethyl acetate. Put one or two filter papers of the same diameter with jar and use it as killing bottle. Temporary killing jars can be prepared by keeping small chips and saw dust soaked in ethyl acetate or blotting paper or cotton soaked in either of the chemicals *viz.* Ether, Benzene, Chloroform, CCl_4 etc.

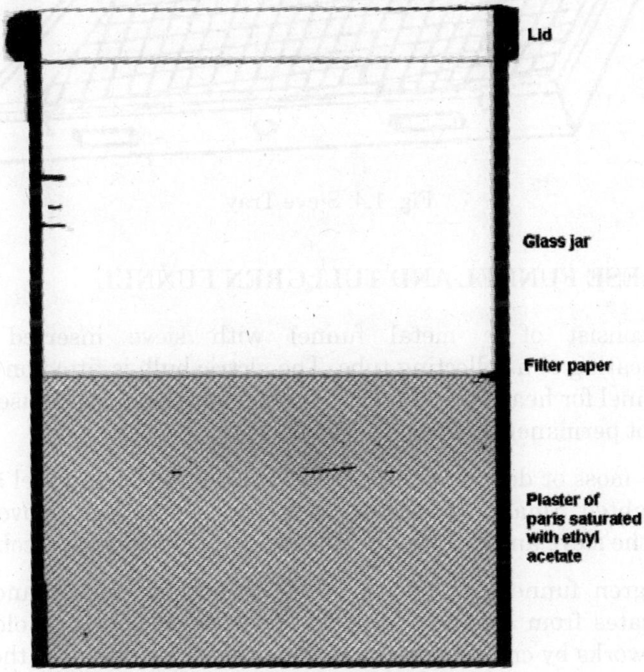

Fig. 1.3. Simple killing bottle

4. COLLECTION VIALS

Homeopathic vials or similar other vials are used to preserve small specimens. Vials with screw plastic caps are preferred. Corning or Borosil tubes are also available in the market but they are very costly.

5. SIEVE TRAY AND ENAMEL TRAY

During preliminary examination, debris and litter etc. is sort out with the help of Sieve tray. The useless parts are discarded through sieve after checking the desired material or simple enamel tray can also be used.

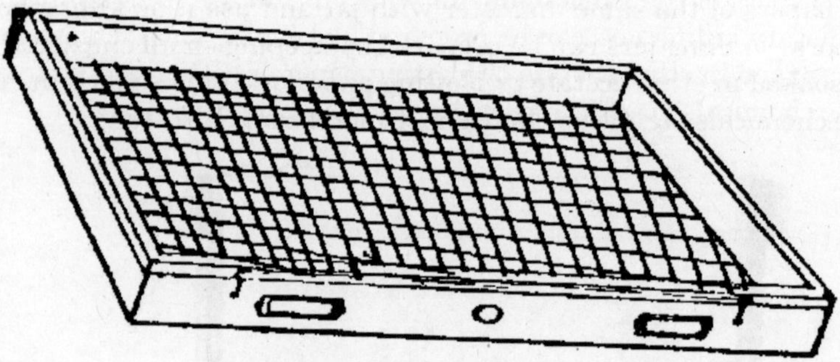

Fig. 1.4. Sieve Tray

6. BERLESE FUNNEL AND TULLGREN FUNNEL

It consist of a metal funnel with sieve inserted inside communicating to a collecting tube. The elctric bulb is fitted on the top of the funnel for heating and light, or table lamp can also be used if the bulb is not permanently fitted.

The moss or debris or litter etc. is poured into the funnel and the bulb is lighted. Smaller insects in order to evade the heat, moves down through the sieve and falls in the collecting tube filled with alcohol.

Tullgren funnel can also be used to extract insects and other invertebrates from a variety of soils, litter, flood debris or old bird's nests. It works by creating warm, dry and light conditions at the top of the funnel which encourages cool, shade and moisture loving invertebrates to move down the funnel away from the light source until they eventually fall out of the bottom of the funnel into a collecting bottle. If you requires live specimens then a lightly moistened filter paper can be placed in the collecting container. It is left in operation for a week or so and inspected daily for the insects collected.

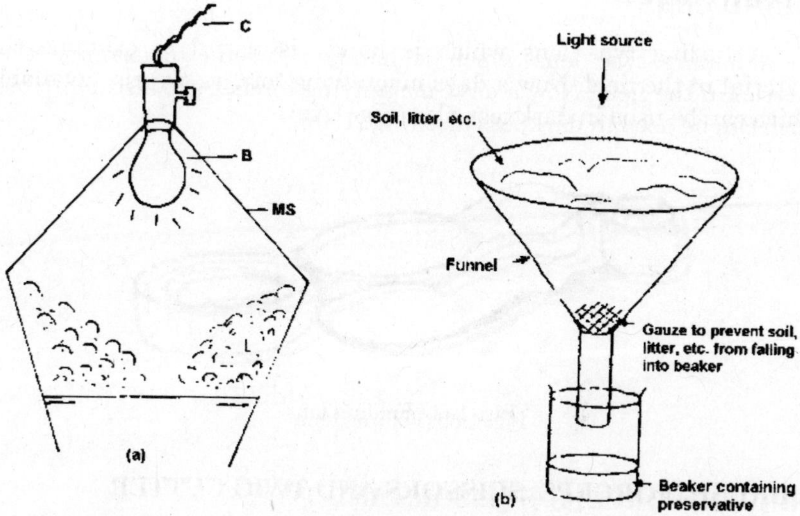

Fig. 1.5. (a) Berless funnel (C-cord, B-Bulb, MS - Metal shade), (b) Tullgren funnel.

7. INSECT BOX

They are made up of plywood 18" x 12" x 4" in size and lined on either one side or both with corksheet covered with white paper. The top can be of glass or wood. In big institutes the collections are kept in glass-toppered drawers kept in insect cabinet (Fig. 1.6a & b).

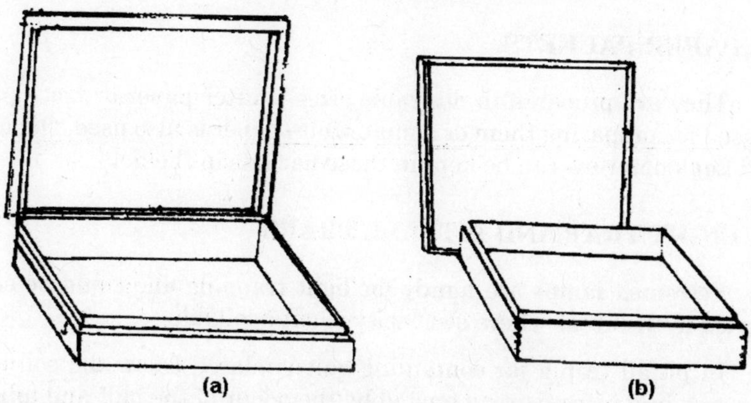

Fig. 1.6. (a) Insect box with lid attached by hinges (b) Insect box with removalbe lid for easy handling of specimens.

7. HAND LENS

A folding type lens which is handy is useful to examine the material in the field. Now a days magnifying torches are also available which can be used in darkness also (Fig. 1.4b).

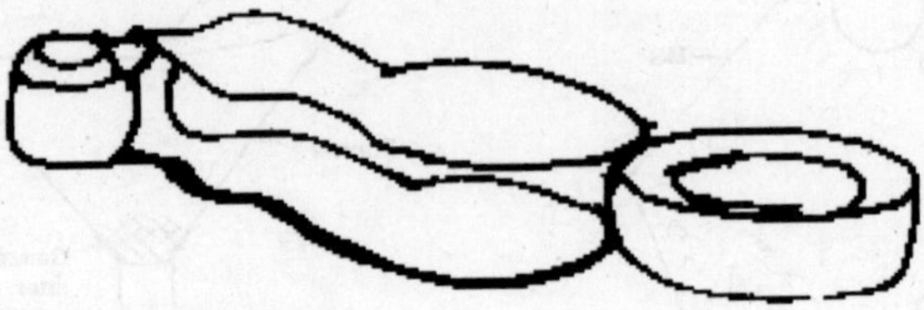

Fig. 1.6b. Folding lens

8. BRUSH, FORCEPS, SCISSORS AND TWIG CUTTER

Soft camel hair brush (number zero (0) or 1) is usually used for hand collections. A sharp forceps with bent ends or straight is used to pick up insects. A twig cutter is used to cut plant parts during collections.

9. AXE, KNIFE AND HAMMER

Termite mounds, bark and rotten logs are accessible by these tools. They are used for breaking and open the mounds, digging out, splitting the wood or loose bark.

10. PAPER PACKETS

They are prepared in desirable sizes. Butter paper or trace paper is used for preparing them or simple white paper is also used. Odonates and Lepidopterans can be kept in these packets in the field.

11. LIGHT TRAP AND PITFALL TRAPS

Petromax lamps are handy for light trapping and could be easily carried in the areas where electricity is not available.

In pitfall trap a jar containing bait is placed below the soil level. The crawling insects are attracted by the odour of the bait and fall into the jar filled with alcohol.

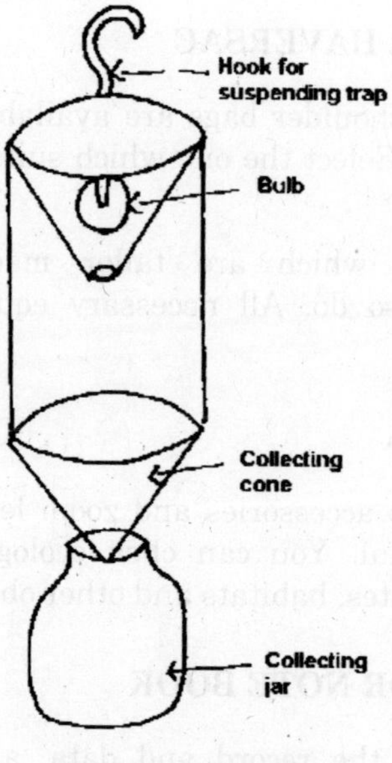

Fig. 1.7a. Light trap

An artificial light is placed adjacent to a white cloth in the field which would attract a number of insects like crickets, grass hoppers, moths, beetles, mantids etc. The insects would rest on the white cloth where they can be put into collecting medium.

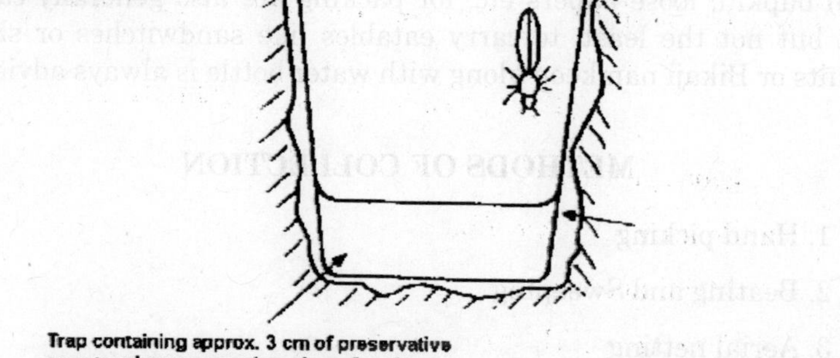

Fig. 1.7b. Pit fall trap

12. CARRY-BAG OR HAVERSAC

Very attractive shoulder bags are available in the market which are water proof also. Select the one which suits your pocket and fulfill requirements.

Coat or jacket which are tailor made according to your specifications can also do. All necessary equipments could best be carried in Haversac.

13. FIELD CAMERA

A set of close up accessories and zoom lens along with any SLR Camera is very useful. You can click ecological conditions, insects community, feeding sites, habitats and other observations.

14. FIELD DIARY OR NOTE BOOK

For keeping all the record and data, a diary or note book is essentially required. It is always desirable to write about the date of collection, place, nearest town and distance from the road or national high way, direction, descripton of the habitat, colouration of the live insects, name of the host plant or animal, associated animals or insects, name of the collector.

15. MISCELLANEOUS

Preservative like 90% alcohol, cotton, killing and preserving fluids, blotting paper, ether, ethyl acetate, extra vials, jars etc are generally kept in the carrying bag. Blade, bandage, soap (phenolic or dettole soap) napkin, loose papers etc. for packing are also generally carried. Last but not the least, to carry eatables like sandwitches or slice or biscuits or Bikaji namkeen along with water bottle is always advisable.

METHODS OF COLLECTION

1. Hand picking
2. Beating and Sweeping
3. Aerial netting
4. Collecting with aspirator 5. Trapping
6. Use of Berlese funnel.

(a) Hand Picking

Soft bodied insects are best collected by hand picking with the help of fine camel brush soaked in alcohol or light weight forceps. Insects like leaf miners, aphids, bark beetles earwigs, beetles, ants, termites etc. are collected by hand picking.

(b) Beating and Sweeping

A long stick and white cloth or tray are required in beating. Insects are easily dislodged from the trees or foliage when they are beated with the help of a stick. The insects are collected on the white cloth spread on the ground.

Sweeping is done with a proper net. A net with 60 cm. handle and 20 cm depth bag is easy to sweep Flying insects are collected by sweeping.

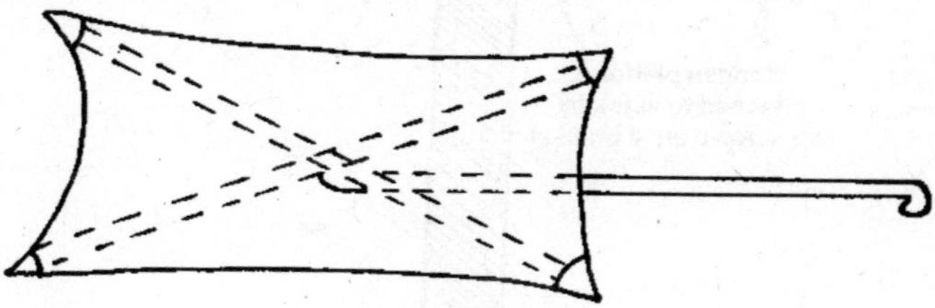

Fig. 1.8. Beating tray

(c) Aerial netting

Odonates, Lepidopterans, Hymenopterans and Dipteran insects are best collected by aerial netting. Do not attempt to get a large number of specimens into your net at one time. The insects collected in net can be transfered into the killing jar by holding the bottom of the net bag vertically up in the air and inserting the jar or bottle into the net.

(d) Collecting with aspirator

Some insects which are minute and delicate and cannot be collected by net are best collected by aspirator. Air is sucked through rubber tubing which draws the insects into the main tube. The lid of the main tube can be removed to emptied the contents into killing bottle or preservative.

(e) Trapping

The insects which otherwise evade attention are collected by traps. Trapping is also useful in the study of migration seasonal abundance, peak population, incidence of pest attack, migration etc. Water traps, flight interception traps, wind traps, light traps, pit fall traps are some common examples.

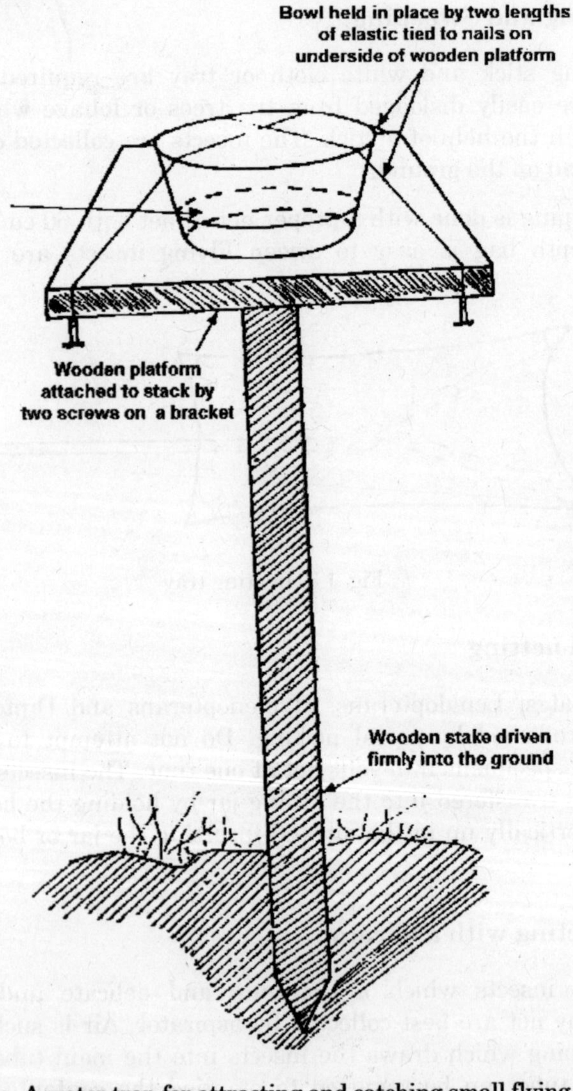

Fig. 1.9a. A water trap for attracting and catching small flying insects.

Fig. 1.9b. Water trap

(f) Water traps

It is used for many flying insects which are attracted to certain colours and caught in coloured water-filled bowls. Yellow bowls are best for flies and Hymenoptera. White colour also attract flies but repells Hymenoptera. Height of the trap influences the species composition of the catch. To catch a wide range of species a number of traps should be used and set at different heights. Catches are heighest when the trap is just above the level of surrounding vegetation when the sampling is done in forest, it is necessary to cover the trap by wide-mesh gauge over trap to prevent leaves from falling in and affecting the attraction of the trap. Trap should be emptied once a week.

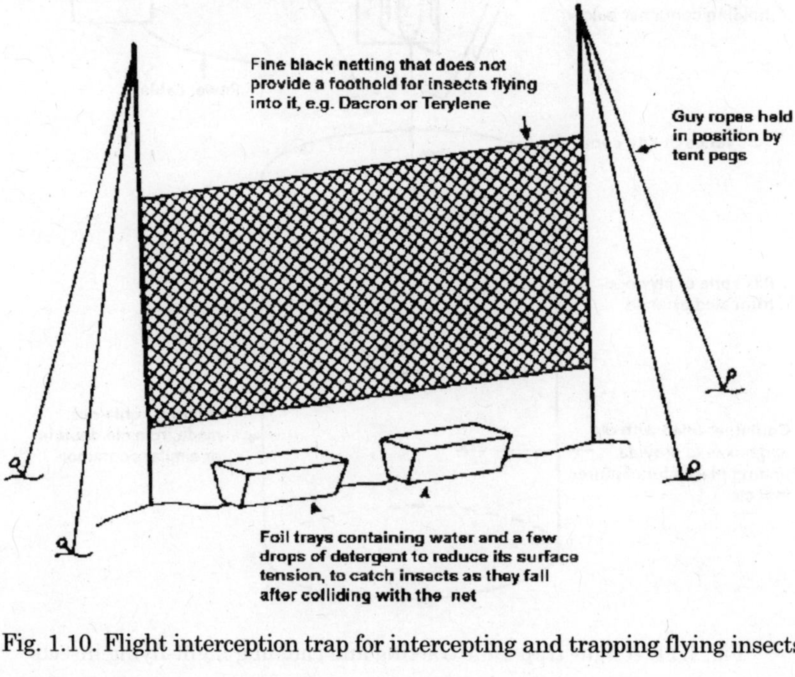

Fig. 1.10. Flight interception trap for intercepting and trapping flying insects

(g) Flight interception traps

Such traps work by blocking flying insects with a screen of fine black netting. Blocked insects then drop down into collecting trays laid beneath the netting. These traps can be sited in areas frequented by large numbers of flying insects such as forest edges or woodland ridges.

(h) Light traps

They are mainly for moths but other night flying insects are also caught. Night flying insects are attracted towards light particularly that at the ultraviolet end of the spectrum. They are either actively caught or encouraged to enter a trap.

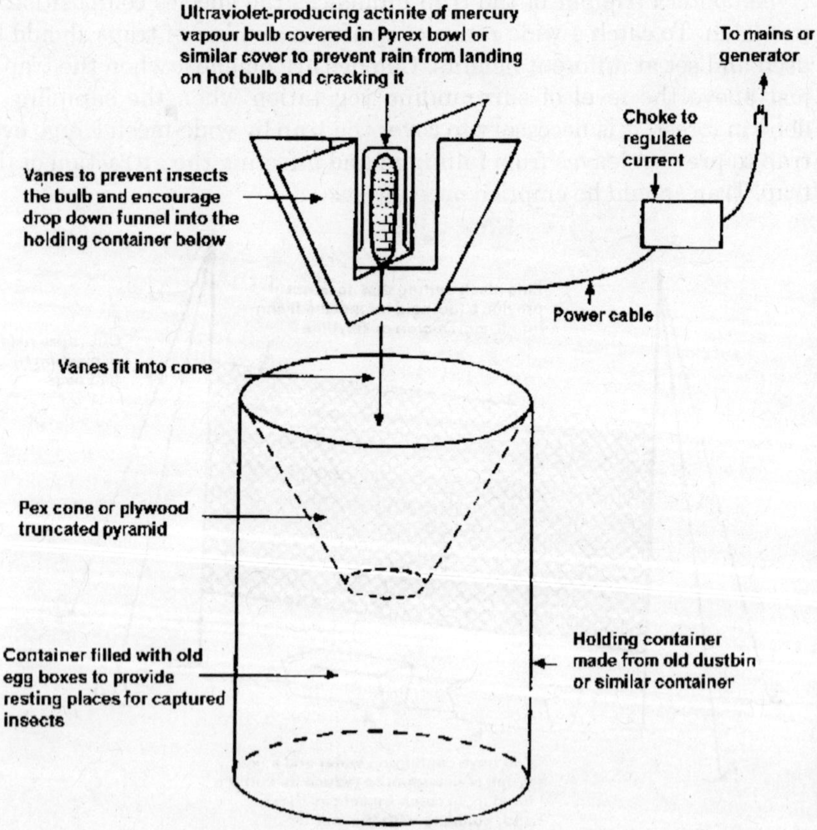

Fig. 1.11. A light trap for attracting and catching night-flying insects

The simplest light trap consists of a light on a cable hanging outside the building. Any bright white or bluish light is suitable although high-pressure mercury vapour bulb is best. The effective-ness of the trap can be enhanced if the lamp is positioned beside a white wall or has a white sheet hung next to it. Alternatively the light can be hung from a tripod above the white sheet. If electricity supply is not available then gas lamp or paraffin vapour can be used instead. Light traps attract most moths on warm overcast nights with little wind, and specially when it is thundery.

Note : Light traps should not be operated during rain because water falling on the hot bulb will crack it.

(i) Aerial attractant traps

Flies can be attracted into containers holding suitable baits and then trapped within these or guided upwards into a collecting bottle. A wide range of baits such as rotting fruit for fruit flies, dung for-dung flies, carrion for carrion-feeding species, carbon dioxide in the form of dry ice wrapped in polythene for slow release for blood-sucking species, fungi, fish, rotting eggs, etc. Butterflies can also be attracted by rotting fruit (over-ripe banana) or dung (Fig. 1.12a & b).

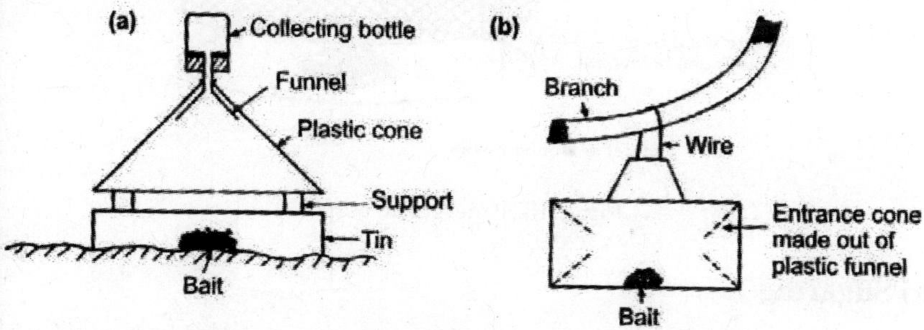

Fig. 1.12. Aerial attractant trap (a) ground based (b) hanging type

(j) Emergence traps

It is mainly used for flies but May-flies and Caddis flies can also be collected. Such traps can be used to trap adult insects emerging from their pupae in water, soil, dead wood, etc. The basic design of these traps consists of a container that fits snugly over the substrate or water surface from where the adult insects emerge. This container usually has an opening to the light at its highest point, which the entrapped insects move towards. Insects caught in traps without collecting bottles can be removed by spraying with 70% alcohol or by using a brush soaked in 70% alcohol.

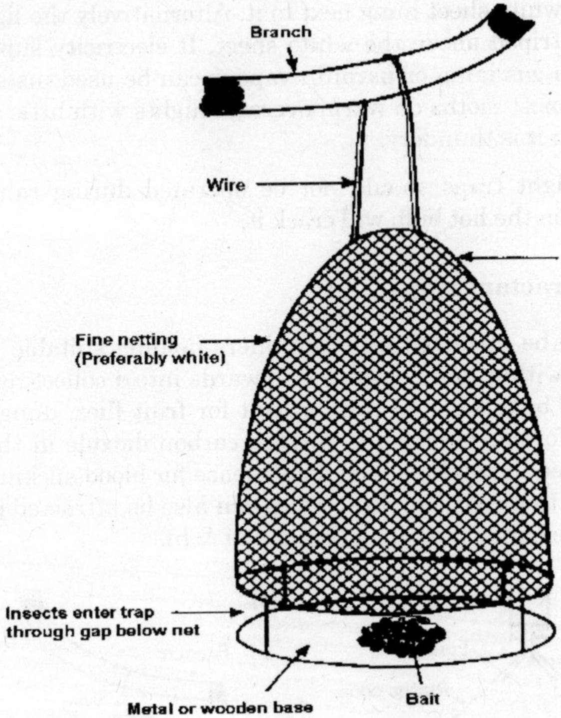

Fig. 1.13. Emergence trap

(k) Sugaring

This method consists of attracting moths to a sugary solution laced with alcohol, which renders them less agile. Sugaring attracts the largest number of moths during warm, overcast weather. The molasses sugar is boiled up with over-ripe bananas or other rotting fruit. Just before use, some alcohol (e.g. beer) should be added and a few drops of amyl acetate to give strong scent to the mixture.

The `sugar' should be applied at dusk on foliage, tree trunks, concrete or wooden posts using a point brush. Sheltered spots are best if it is windy. Sugaring requires the constant presence of the recorder so that, as soon as the moth is attracted it can be separated. Sugaring often attracts those moths that do not usually come to light.

(l) Use of Berlese funnel

Insects inhabitating sub soil, leaf litter or inside the soil are collected and carried to laboratory where they are subjected to Berlese funnel (Several modifications of this device are available). The simplest device is to keep a funnel on the bottle which is fitted with clamp and stand. Over the funnel lighted a lamp. The soil is kept in the funnel on the gauge fitted there in and the bulb is lighted. The insects evades the light and heat and are collected in the bottle containing alcohol (see Fig. 1.5a).

TYPICAL ABODES OF INSECTS

Topography and climatological conditions of Indian sub continents are vast. North-east and north west parts of India offer fairly good collections. Given below are some typical habitats of insects from where they can be collected.

Plant parts

Root, stem, leaves, flowers and fruits. A number of insects like aphids, beetles, moth and butterflies, bugs, leaf miners, thrips, bees, fruit flies etc can be collected from these parts.

Algae, fungi, mosses and lichens

Spring tails and beetles can be collected.

Leaf litter and under stones

Earwigs, ground beetles, collembola etc. can be collected.

Rotten wooden logs

Beetles, termites and Hymenopterans can be collected.

Light (Street light, Porch light)

Beetles, bugs, moths, grasshoppers and mosquito and flies can be collected.

Dung and carrion

Carrion beetles, dung rollers, Dipteran larvae can be collected.

Graneries, ware houses

Stored grain pests can be collected.

Tree bark

Bark beetles can be collected.

Table 1. Collection and preservation

Order	Collection	Killing	Preservation
Protura Diplura Collembola	Berlese funnel	95% alcohol	90% alcohol (permanent slide)
Thysanura	Camel hair brush soaked in alcohol	Carbon tetra chloride	
Blattaria Mantodea	Hand picking/ sweeping	Benzene, ether, ethylacetate or cyanide vapours	Pinned
Dermaptera	Forceps	ethyl acetate	alcohol
Plecoptera	Netting, Sweeping	95% alcohol	90% alcohol
Isoptera	Forceps/brush soaked in alcohol	95% alcohol	90% alcohol (permanent slide)
Embioptera	Brush/Forceps	95% alcohol	Adult pinned or 90% alcohol
Ephemer-optera	Brush soaked in alcohol light	95% alcohol	90% alcohol (permanent slide)
Odonata	Aerial netting	ethyl acetate (nymph 95% alcohol)	Pinned
Psocoptera	Aspirator brush soaked in alcohol	—	alcohol (permanent slide)
Anoplura	Brush soaked in alcohol or forceps	95% alcohol	90% alcohol (permanent slide)
Thysanoptera	Brush soaked in alcohol beating Berlese funnel	95% alcohol	90% alcohol (permanent slide)
Hemiptera	Sweeping aquatic net brush soaked alcohol light trap aspirator	Ethyl acetate 95% alcohol	Pinned alcohol Berlese medium
Neuroptera	Net sweeping light and hand picking	ethyl acetate	Pinned
Mecoptera	Aerial netting light trap	ethyl acetate 95% alcohol	Pinned 90% alcohol

Trichoptera	Light trap Aerial net	ethyl acetate 95% alcohol	Pinned 90% alcohol
Lepidoptera	Aerial netting sugar bait rearing larvae	ethyl acetate	Pinned
Coleoptera	Sweeping Hand picking beating aquatic-net	ethyl acetate	Pinned 90% alcohol
Hymenoptera	Rearing beating sweeping light trap	ethyl acetate	Pinned 70% alcohol permanent slide
Siphonoptera	From the host's body	95% alcohol	90% alcohol permanent slide after cleaning in 10% KoH
Diptera	Sweeping aerial netting aspirator Berlese funnel rotten fruits debris and mud meat and dung	95% alcohol ethyl acetate	90% alcohol pinned permanent slide

PRESERVATION

DRY PRESERVATION

Dry preservation facilitates storage and production of attractive specimens with most of the colouration well preserved. In drying specimens are arranged in such a way that all parts of the body are visible for inspection under microscope and protected. Dried insects are set. They are brittle and very fragile therefore requires utmost care. Given below are the steps followed in dry preservation.

1. Temporary storage

Some insects like Butterflies and Dragon flies can be kept in paper envelops if the immediate mounting is not possible. Other insects can be stored in card board with cotton bed and they can be relaxed for mounting at convenient times.

2. Relaxing

A relaxing box consist of a layer of absorbent cotton, rubber sponge or sand saturated with water at the bottom of the box. The insects in butter paper envelops are kept taking care of avoiding the direct contact of the insects with water. *Paradichloro benzene* or *napthalene balls* can be kept in the box to prevent fungal infections and ward off ants and other insects.

The insects are allowed for 24 hrs in the relaxing box or depend on the size of the insect. The relaxing box has high humidity all the time. Insects thus relaxed are examined taxonomically. Relaxing helps to keep the bodies, appendeges movable and flexible.

3. Cleaning

Any dirt or adhered material to the body of the insect is cleaned with the help of camel hair brush dipped in ammonia and water or only water.

Greasy specimens can be dipped in fat solvent like kerosene. Before mounting they should be dried.

4. Pinning

The common method of dry preservation is pinning. Special non corrosive pins are either directly or on secondary mounts are inserted. Each order has a specific pinning position so that the insect is properly oriented and balance and all characters necessary for identification are visible. It is necessary to choose a pin of correct size and number. Insect pins vary in length and diameter and are numbered 000, 00, 0, 1, 2, 3, 4, 5 etc. 000 is very slender and the 5 is most stout. Pins in the size 0-3 are generally used no 4 & 5 are approximate 37-39 mm. length and are used for large bodied insects.

Where to pin

Hymenoptera

Bees and wasps are pinned through thorax, to the right of the middle line, between or little behind the base of fore wings.

Coleoptera

Beetles are pinned through the right elytra, near the front margin, near middle line.

Hemiptera

Bugs are pinned through scutellum to the right of the middle line.

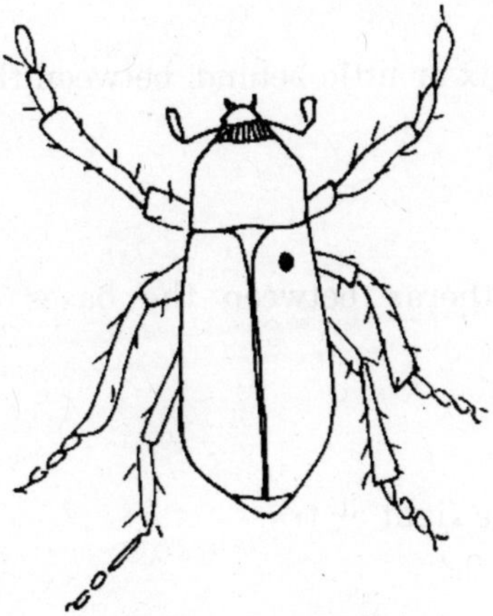

Fig. 1.14. Coleoptera

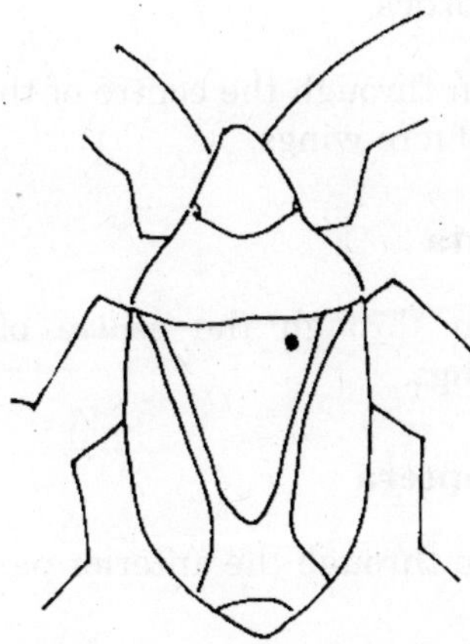

Fig. 1.14. Hemiptera

Orthoptera

Grass-hoppers are pinned through the back part of prothorax, to the right of the middle line.

Mantodea

Mantids are pinned through the centre of meta- thorax.

Blattodea

Cockroaches are pinned through the centre of meta-thorax.

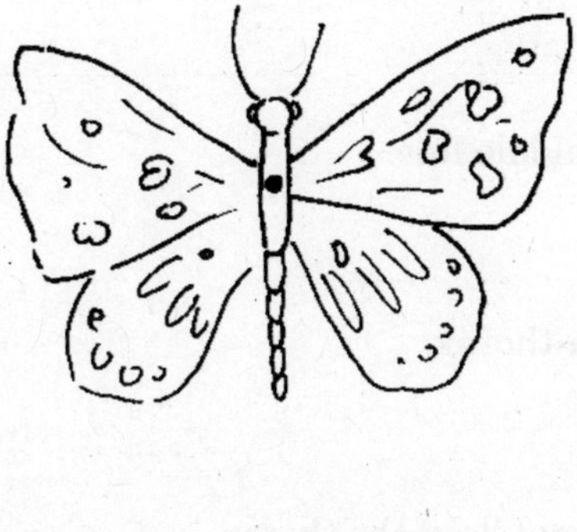

Fig. 1.14. Lepidoptera

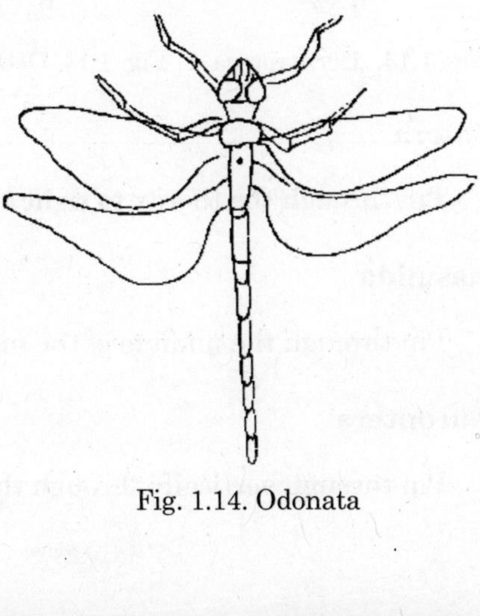

Fig. 1.14. Odonata

Lepidotera

Pin through the centre of the thorax or little behind, between the bases of fore-wings.

Odonata

Pin through the centre of the thorax between the bases of fore-wings.

Dermaptera

Pin through the anterior part of the right elytra.

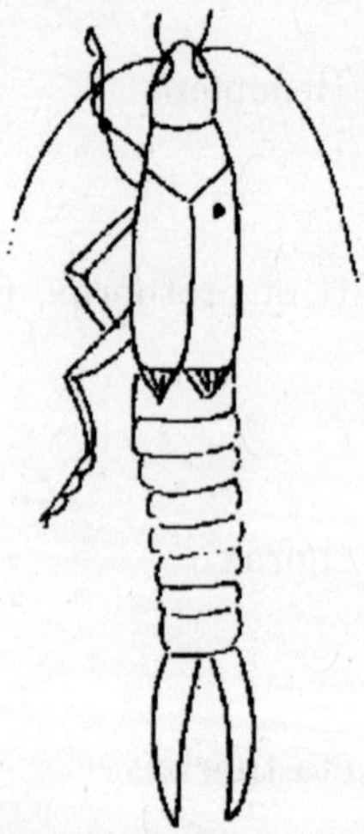

Fig. 1.14. Dermaptera

Fig. 1.14. Orthoptera

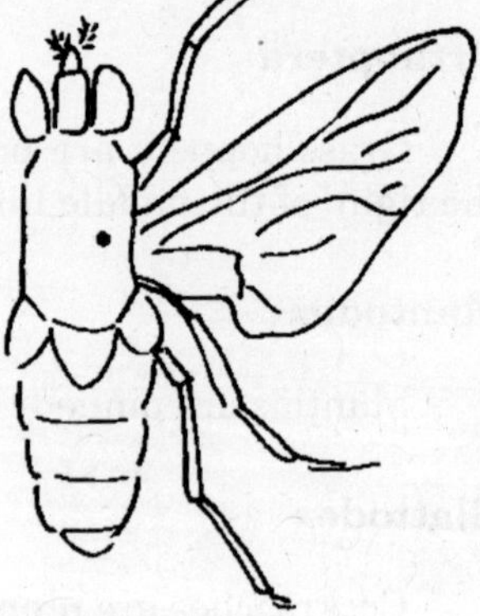

Fig. 1.14. Diptera

Diptera

Pin through the thorax to right of middle line.

Phasmida

Pin through the midline of the meta-thorax.

Neuroptera

Pin through vertically through the middle of the thorax.

Mecoptera and Trichoptera

Pin through the middle of the thorax.

5. Spreading

A spreading board consist of two flat parallel pieces of soft wood with an inner groove lined by cork. Wings are arranged and spreaded on the board. Narrow straps of papers are used to hold the wings. The spreading board with insects are kept in dry airy place. After the insect dries, they are transferred to insect box.

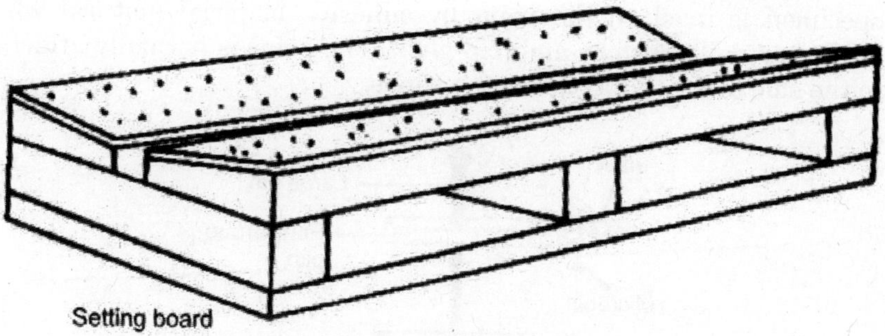

Fig. 1.15a. Setting board.

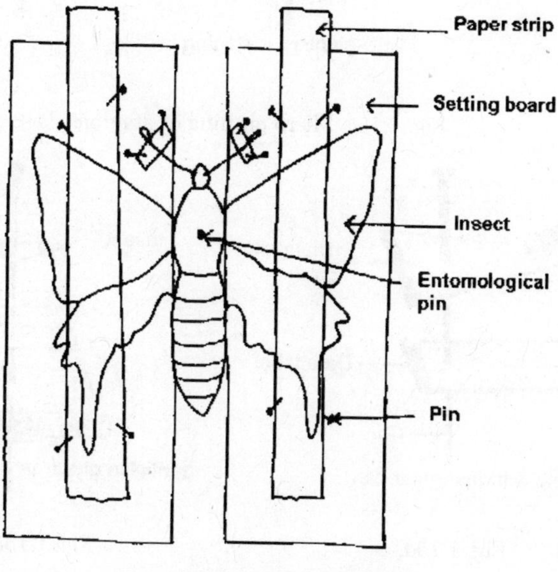

Fig. 1.15b. Setting of butterfly.

6. Double mounting

Staging or double mounting is useful for small insects. Small delicate specimens can be pinned with a minute, fine pin on a piece of a pith (cork), through other end, a longer pin is inserted. The labels are attached to this long pin.

7. Point mounting

Small insect can also be mounted on small broadly triangular piece of stout paper or card. A pointed triangular flag is prepared from the card and a entomological pin is inserted on its broad end. The specimen is fixed on the point by adhesive material (natural white shellac or nail polish or ambroid glue). The point is normally attached to the side of the thorax below the wings.

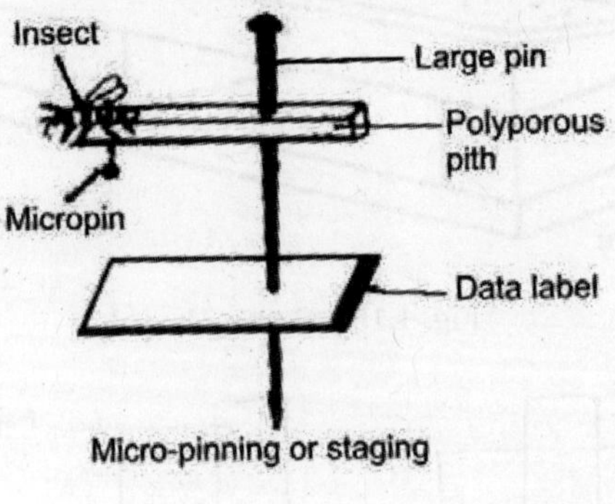

Fig. 1.15c. Micro pinning or staging

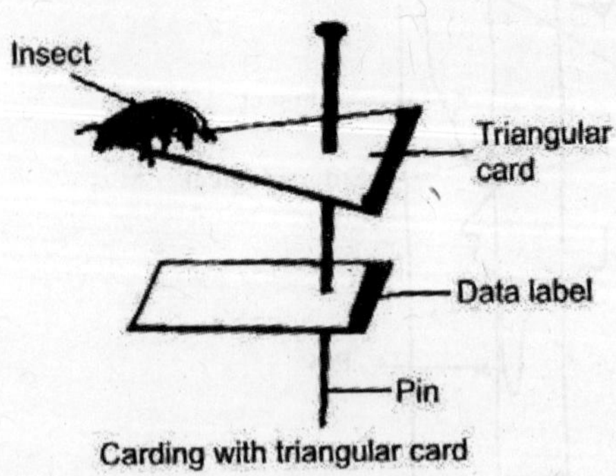

Fig. 1.15d.

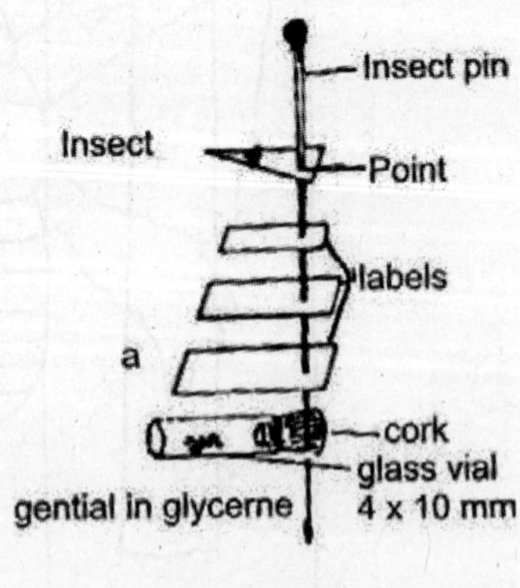

Fig. 1.15d[1].

8. Glassine preparations

Antlions (Neuroptera) and Odonates can be preserved in glassine envelops. The fresh insects collected from the field are arranged properly with its appendages and wings. All characters are clearly visible from outside. Put a label and do not seal the envelop till it dries completely.

9. Drying

Drying of the spreaded insects is a must before storage. It should be dried quickly. Allow it to dry at room temperature and use a drying cabinet.

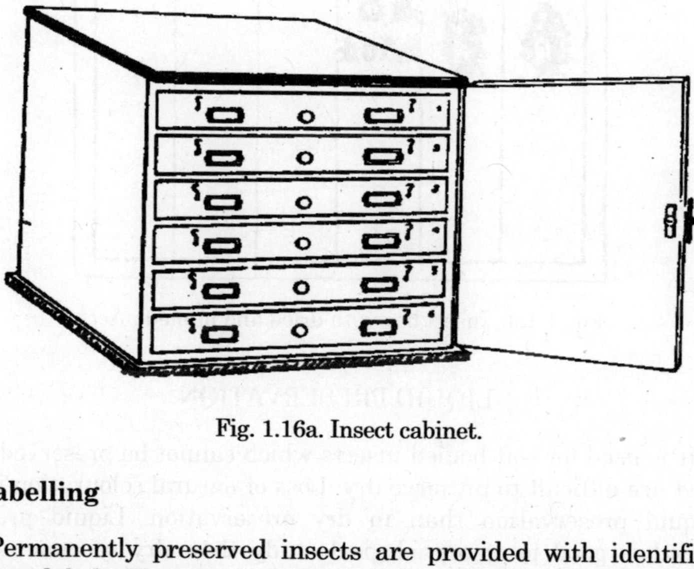

Fig. 1.16a. Insect cabinet.

10. Labelling

Permanently preserved insects are provided with identification and data labels. They are written on good quality paper with India ink or Chinese ink or they can be printed. Data label should contain the date of collection, locality, name of collector, whereas identification label should contain the scientific-name with author,

Species	-----------	Date	------------
Genus	-----------	Locality	------------
Family	-----------	Name of Collector	-------------
Order	----------		

11. Storage

The specimens are subject to damage by various pests. They can be ruined by dust, light or dampness therefore a proper care and maintenance is required. The specimens are stored in air tight boxes. The boxes should contain *paradichloro benzene or a mixture of napthalene and paradichloro benzene or napthalene balls alone.* It should not be exposed to light as the specimens will loose its natural colouration.

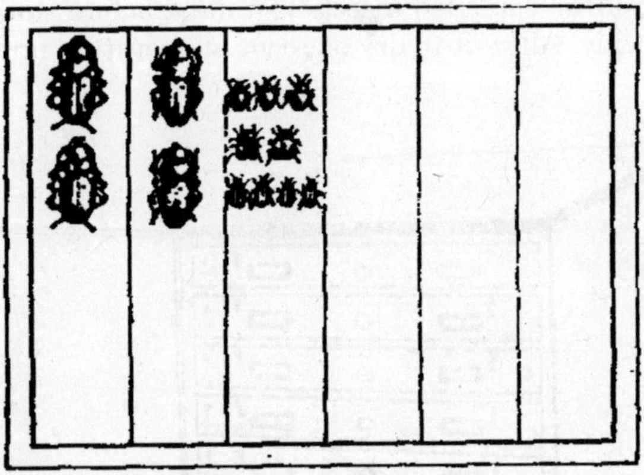

Fig. 1.16b. Insect box with dried and pinned insects

LIQUID PRESERVATION

It is used for soft bodied insects which cannot be preserved dry or atleast are difficult to preserve dry. Loss of natural colouration is more in liquid preservation than in dry preservation. Liquid preserved material is good for morphological study than dry specimens. Good results are obtained with proper fluids in which less shrinkage and distortion of the shape takes place.

The most widely used fluid preservative is 75-90% ethyl alcohol 2% glycerine can also be added to prevent the drying of specimens (when alcohol evaporates) 2-3 change are necessary in the initial preservative because the dilution of the preservative fluid takes place by body fluids. Solution recommended for the detailed study of internal anatomy is —

Glacial acetic acid — 1 part

Ethyl alcohol — 8 part

Distilled water — 5 part

Glycerine — 1 part

Dissecting material is preserved in *Pampils fluid* : (glacid acetic acid 2 parts, formaldelyde (4%) 6 parts, distilled water 30 parts, alcohol 95% 15 parts). Solution of Chloral hydrate is also used.

How to care a liquid preserved material

The smaller forms are mounted on a slide permanently. Large soft bodied insects are preserved is 75% ethyl alcohol. A few changes of alcohol are must as and when the time permits. The material is transferred to new vials with fresh alcohol containing required data as name of locality, lot number, collector's name etc. Homeopathic vials are best suited. More than one species are not put into the same vial. Use separate vials for different species. If the alcohol evaporates, 4% formalin can also be added to the alcohol. If the alcohol tends to evaporate it is advisable to put the vials into a big wide mouth jar containing alcohol and the alcohol can be added time to time in this jar.

How to do dry mounting from liquid preserved material

Minute hard bodied insects can be dry mounted back from liquid preservatives. Follow these steps -

I. Transfer the liquid preserved specimen to 100% ethanol for several hours.

II. Transfer to benzol or kerosene for several hours.

III. Pin directly or double mount the material.

STUDENT COLLECTION

Every student of entomology should collect the local forms. The required insect collection will consist of atleast 200 specimens representing atleast 45-50 families from atleast 18 orders. Students are advised to collect immature stages also. Nymphs of winged and wingless forms, naids, different types of larvae, different types of pupae, and egg. Student are also required to submit collection of different types of nests (in Hymenoptera) and empty pupal shells in Lepidoptera, galls and damaged articles (in case of termites). Life cycle of different insects can also be presented.

CHAPTER - 2

CLASSIFICATION OF INSECTS

INSECTA

Sub-Class

APTERYGOTA

ORDER

1. Thysanura
2. Diplura
3. Protura
4. Collembola

PTERYGOTA

DIVISION

EXOPTERYGOTA (Hemimetabola)

ORDER

1. Ephemeroptera
2. Odonata
3. Plecoptera
4. Grylloblattodea
5. Orthoptera
6. Phasmida
7. Dermaptera
8. Embioptera
9. Dictyoptera
10. Isoptera
11. Zoraptera
12. Psocoptera
13. Mallophaga
14. Siphunculata
15. Hemiptera
16. Thysanoptera

ENDOPTERYGOTA (Holometabola)

ORDER

1. Neuroptera
2. Mecoptera
3. Lepidoptera
4. Trichoptera
5. Diptera
6. Siphonaptera
7. Hymenoptera
8. Coleoptea
9. Strepsiptera

Total 29 orders

Note : The Entomological Society of Republic of China in 1994 has published modern classification of Insecta including 31 orders.

1. DERIVATION OF NAMES OF ORDERS

1. Thysanura (Thysanos, fringe; ura-tail)
2. Diplura (= Aptera) (Di-two, ura-tail)
3. Protura (Pro-first; oura-tail)
4. Collembola (Colla-glue, embolon-boll)
5. Ephemeroptera (Ephemeros -lasting a day, ptera-wing)
6. Odonata (Odoun-tooth)
7. Plecoptera (Plecos-plaited, ptera-wing)
8. Grylloblattodea (Gryllus -a cricket; blatta- a cockroach)
9. Orthoptera (Orthos-straight; ptera-wing)
10. Phasmida (a specter, an apparition) walking sticks.
11. Dermaptera (Derma-skin; ptera-wing)
12. Embioptera (= Embidina) (Embios-lively; ptera-wing)
13. Dictyoptera (Dictyon-net; ptera-wing)
14. Isoptera (Isos-equal/similar, ptera-wing)
15. Zoraptera (Zoros-purely; aptera wingless)
16. Psocoptera (=Corrodentia) Corrodere - to gnaw)
17. Mallophaga (Mallos- wool, phagein - to eat)
18. Siphunculata/Anoplura (Anoplos-unarmed, oura-tail)
19. Hemiptera (Hemi-half; ptera-wing)
20. Thysanoptera (= Physapoda) (Thysanos-fringe, ptera-wing)
21. Neuroptera (Neuron-nerve; ptera-wing)
22. Mecoptera (Mecos-length, ptera-wing)
23. Lepidoptera (Lepidos-scale; ptera-wing)
24. Trichoptera (Trichos-hair; ptera-wing)
25. Diptera (Di-two; ptera-wing)

26. Siphonaptera (Siphon-siphon; aptera-wingless)
27. Hymenoptera (Hymen-membrane; ptera-wing)
28. Coleoptera (Coleos-sheath; ptera-wing)
29. Strepsiptera (Strepsis-twisted; ptera-wing)

CHAPTER - 3

SURVEY OF REPRESENTATIVE INSECT ORDERS

ORDER-THYSANURA (Bristle tails)

Example : Silver fish (*Lepisma*)

1. Mouthparts are *entognathous*. They are adapted for *biting and chewing*.
2. Antennae many segmented.
3. Tarsi 2-5 segmented provided with a pair of claw.
4. Many segmented paired cerci, median caudal filament is present.
5. Abdomen 11 segmented (last segment is much reduced).
6. Lepisma is a soft flat bodied insect. Integument is covered with scales.
7. Development is *epimorphic* i.e. the number of postcephalic segments increases by proliferation in the hind part of the body following the first embryonic stages, but becomes fixed during the embryonic period. Thus the insect hatches from the egg with the adult number of segments.

Habit and Habitat

They are found hidden in cracks and crevices in the buildings, understones, in rotting wood. They are fast runners and active in darkness.

Food

Silver fish feeds on a large variety of materials such as starched clothes, rayon fabrics, binding of books, papers, book labels. They are probably omnivorous.

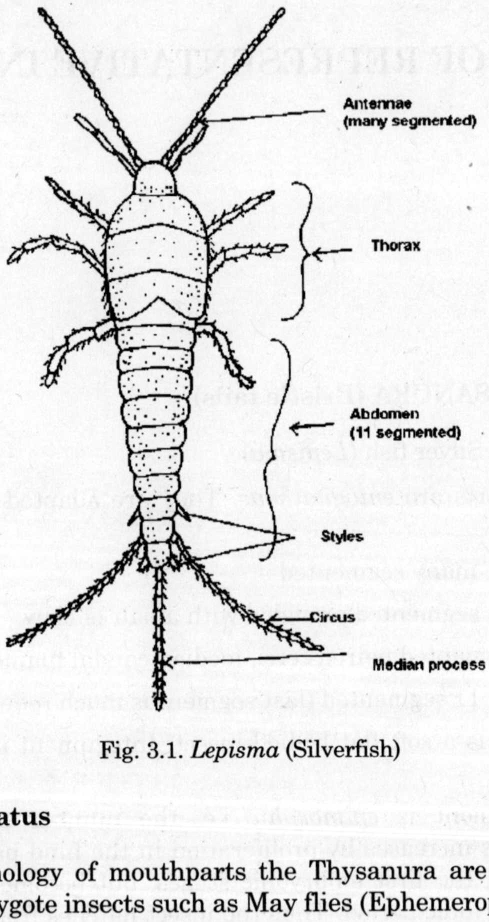

Fig. 3.1. *Lepisma* (Silverfish)

Biological status

In morphology of mouthparts the Thysanura are comparable to primitive pterygote insects such as May flies (Ephemeroptera).

Economic status

They cause considerable damage to books, card files, rayon fabrics, stocks of paper, starched clothings etc. Other articles containing glue are also attacked.

Collection: (i) Preserve in 70% alcohol or prepare a permanent slide; (ii) Easily collected from household.

ORDER - DIPLURA (Double tail)

Example : *Campodea, Japyx*

1. Mouthparts *entognathous* (not visible externally), the nature of mouthparts is *chewing type*.
2. Antennae many segmented.
3. Tarsi 1- segmented provided with a pair of claw.
4. Cerci of variable form.
5. Ametabolous insects.
6. Development is *epimorphic*.

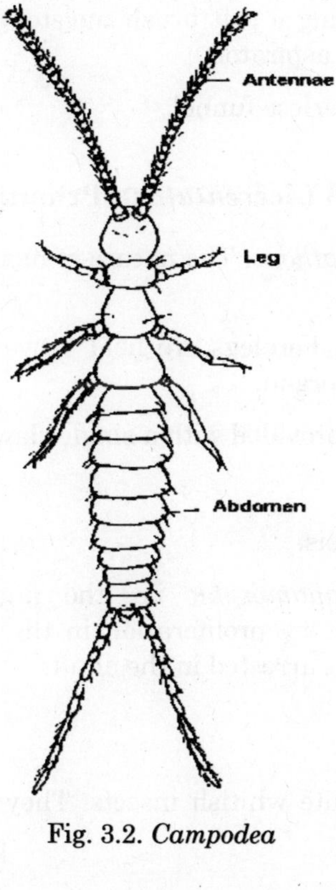

Fig. 3.2. *Campodea*

Habit and Habitat

They are found in damp places in the soil, under barks, under stones or logs, in rotting wood, in debris etc. They are darkness loving.

Food

They feed on decaying vegetation but some are predatroy, use their forceps like cerci to capture the prey.

Biological status

1. They are of considerable phylogenetic significance due to their resemblance to *Symphyta* (Imms, 1936).
2. Japygidae is unique possessing four pair of spiracles (2 pair each on meso and meta thorax).

No economic importance.

Collection (i) Preserve in alcohol or prepare permanent slide. They can be collected by using a soft brush moistened in 70% alcohol, by a pair of forceps and aspirator.

(ii) Extract through Berlese funnel.

ORDER - PROTURA (*Acerentulus*) (Primitive tails)

1. Mouthparts *entognathous*. The nature of mouthparts is *piercing* and *sucking type.*
2. Antennae reduced. Forelegs are held forward like antennae. They are used as tactile organ.
3. Tarsi 1 segmented provided with a single claw.
4. Cerci absent.
5. Slight metamorphosis.
6. Development is *anamorphic* i.e. the number of post cepahlic segments increases by proliferation in the hind part of the body. This proliferation is arrested in the adult.

Habit and Habitat

Protura are minute whitish insects. They occur in most soil and leaf litter.

Food

It constitutes the juices of rotting vegetation and fungi.

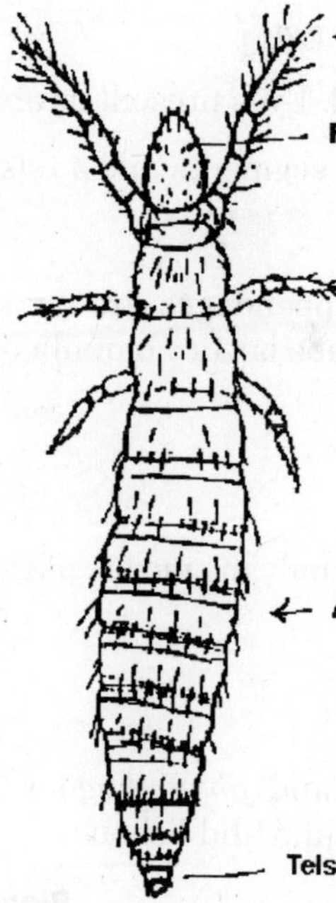

Fig. 3.3. A *Proturan* insect

Biological status

They are unique among insects in -

(i) Antennae absent. Eyes also absent.

(ii) Pseudocelli present.

(iii) Cerci absent.

(iv) Abdomen with 11 segments and well developed telson.

No economic importance.

Collection : Preserve is alcohol or prepare permanent slide.

(They are not often seen owing to their small size. Information regarding the Protura of India are very scanty). They occur between 5 cm and 20 cm deep inside the soil.

ORDER-COLLEMBOLA (Spring tails)

Examples: Tomocerus, Sminthurinus, Sminthurus, Entomobrya.

1. Mouthparts are of *biting type*.
2. Antennae 4-6 segmented. Eyes present or absent.
3. Abdomen composed of 5 segments and a telson.
4. Tarsi and tibia fused.
5. Pregenital abdominal appendages are on 1-3 segment and 4 or 5th segment (Ventral tube on 1st, hamula on 3rd, furca on 4th or 5th).

Habit and Habitat

They are commonly found in moist places; decaying organic matter, under bark of trees.

Food

They are *saprophagous* and *phytophagous*. They feed on pollen grains, fungal spores or mycelium and lichens.

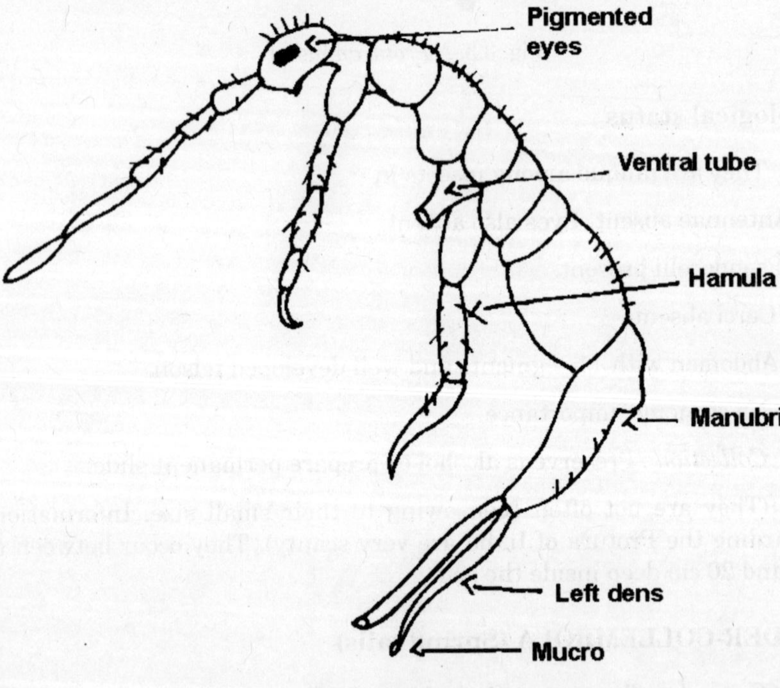

Fig. 3.4. Springtail

Biological Status

It exhibits *Protomorphous* development i.e. at no stage in development, the number of abdominal segments is more than 5+1. It is unique in insecta.

Economic status

Certain species are injurious to mushrooms. They also cause considerable damage to greenhouse gardens. They are also injurious to germinating seeds.

Collection

They can be collected by means of small brush soaked in alcohol. Berlese funnel is most appropriate. They can be preserved in 70 or 90% alcohol. Fresh specimens can be mounted in chloral hydrate.

ORDER - EPHEMEROPTERA (May flies)

Example - Ephemera

1. Mouthparts atrophied in adults.
2. Wings non flexing and held vertically at rest. They are membranous, fan shaped and with reticular venation. Shape triangular.
3. Abdomen terminates into long cerci and median caudal filament which is unique in insects except Thysanura.
4. Development is hemimetabolous. The campodeiform nymphs live in fresh water (often called *naids*). They respire by abdominal tracheal gills which are plumose or lamellate. Adult soft bodied and delicate.

Habit and Habitat

They are familiar on the margins of lakes, streams, rivers and ponds. Adults do not feed. They are attracted to electric lights during night particularly is rainy season.

Food

The nymphs feeds on algae and fragments of plants.

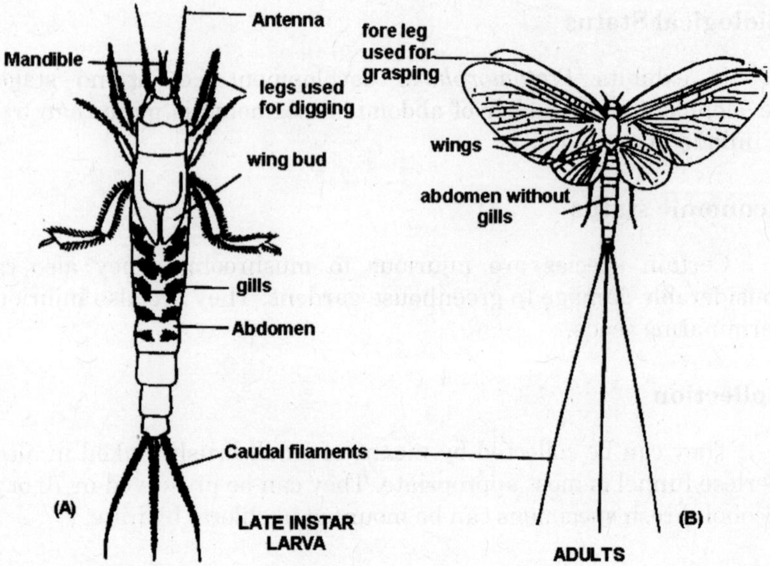

Fig. 3.5. May fly.

Biological status

Adult May flies are *ephemeral* i.e. they live for a day or two. Nymphs metamorphose into winged *sub-imago* which under go one more moult to attain imago or adults condition. This is a unique feature among the insects.

Collection

They can be collected by brush soaked in alcohol. Preserve in 70% alcohol. Permanent slide can also be prepared. They can also be collected by beating vegetation over a paper/blotting sheet. Light source near water body attracts adults. Larvae can be collected by water nets.

ORDER - ODONATA (Dragon flies and damsel flies)

1. Wings possess *nodus*, *pterostigma* and *arculus*. *Nodus* is a thick cross vein at the anterior margin of wing. *Pterostigma* is a spot like structure near the tip of the wing. *Arculus* is a cross vein between radius and cubitus.

2. Two pair of wings are non flexing i.e. they do not fold over the abdomen and remain vertical in Damsel flies and held straight in Dragon flies when the insect is resting.
3. Male copulatory organs are unique in the insects being situated on 2nd and 3rd abdominal sterna.

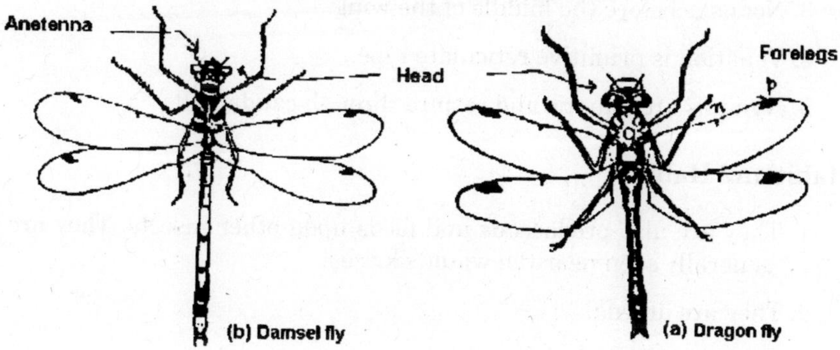

Fig. 3.6a & b.

Sub order - Anisoptera (Dragon fly)

1. Mouthparts are of *biting* type.
2. Antennae (*setaceous*) very small and the eyes are very long.
3. Unequal wings. Hind pair broadened basally and held horizontally at rest. venation reticulate.
4. Nymphs aquatic (*naids*) and respire through rectal gills or caudal gills.

Habit and Habitat

1. They are essentially sun loving found in the proximity of water. Most of the times they are on wings. They rest on plants, fences and other objects.
2. They are *Predaceous*. The labium of nymphs is modified into an elaborate, prehensile organ called *mask*.

Food

Mosquitoes and other insects.

Sub order - Zygoptera (Damsel fly)

1. Fore and hind pair of wings are similar in form and venation, more or less petiolate basally and at rest remains vertically above abdomen.
2. Eyes are widely separated by a space.
3. Nodus is before the middle of the wing.
4. Venation is primitive reticulate type.
5. Nymphs are aquatic and respire through caudal gills.

Habit and Habitat

1. They are also predaceous and feeds upon other insects. They are generally seen near the water sources.
2. They are diurnal.

Collection

Collect through hand net by sweeping and pin in the box. Nymphs can be preserved in alcohol and are collected by water nets amongst rooted vegetation, under pebbles and by scooping and washing bottom mud.

ORDER - PLECOPTERA (Stone flies)

1. Mouthparts *biting* and *chewing* type.
2. Antennae long (with 25-100 segments) and setaceous.
3. Compond eyes small.
4. Thoracic segments nearly equal.
5. Wings membranous, hyaline, vestigial or absent in males of several species. They are folded backwards flat over the abdomen when at rest.
6. Cerci usually long.

Habit and Habitat

1. Stone flies breeds in cold streams.

2. They are cryptically coloured and slow erratic fliers.
3. Nymphs hides under stones or other debris in water.
4. Naids are predaceous on May fly larvae.

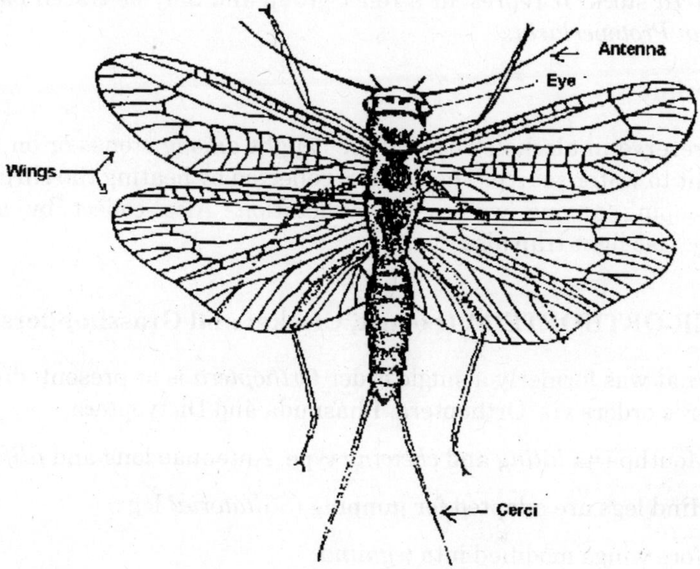

Fig. 3.7a. Plecoptera (Stone fly)

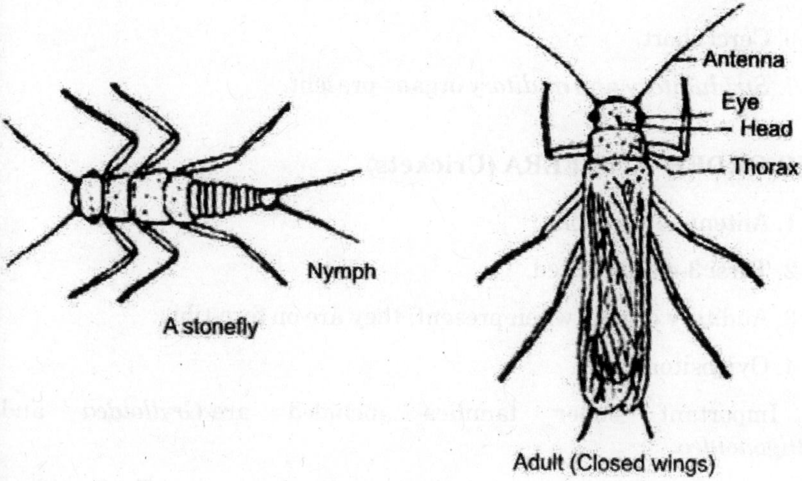

Fig. 3.7b.

Biological Status

The order is neither closely related to *Ephemeroptera* nor to the Odonata, but is probably derived from *Protoblattoidea* or *Proto-orthoptera* stock. It represent a relict group and may be traced back to *Permian Protoperlaria.*

Collection

Preserve in alcohol. They can be caught among stones or on posts adjacent to water source. They can be collected by beating the shrubs or by sweeping among water side vegetation. Also collect by aerial netting. Use light trap also.

ORDER-ORTHOPTERA (Locust, Cricket and Grasshoppers)

What was formerly a single order *Orthoptera* is at present divided into three orders viz. Orthoptera, Phasmida and Dictyoptera.

1. Mouthparts *biting* and *chewing* type. Antennae long and *filiform.*
2. Hind legs are adapted for jumping (*Saltatorial* legs)
3. Fore wings modified into *tegmina.*
4. Tarsi 3-4 segmented.
5. Ovipositor well developed.
6. Cerci short.
7. *Stridulatory* and *auditory* organs present.

SUB-ORDER- ENSIFERA (Crickets)

1. Antennae very long.
2. Tarsi 3-4 segmented.
3. Auditory organs when present, they are on fore-tibia.
4. Ovipositor long.

Important super families included are-*Grylloidea* and *Tettigonoidea.*

Important families are *Gryllidae, Gryllotalpidae and Tettigonidae.*

SUB-ORDER-CAELIFERA (Grasshoppers)

1. Antennae short.
2. Tarsi not more than 3 segmented.
3. Auditory organ when present, is on 1st abdominal segment.
4. Ovipositor short.

Important super families included are *Acridoidea* and *Tridactyloidea.*

Important families are *Acrididae, Tetrigidae* and *Tridactylidae.*

Schistocerca gregaria **(Desert locust)** Fig. 3.8 (c)

Habit and Habitat

Locust is herbivorous and found in tropical regions of the world. It devours and devastate whatever greenery comes its way.

Characters of the genus

1. Body divided into head, thorax and abdomen.
2. The mouth parts are *biting* and *chewing* type.
3. Hind legs are saltatorial (adapted for jumping)

Biological peculiarity

Locust exists in two phases i.e. solitary phase and gregarious phase. It changes from-one phase to another through phase transiens nymphs of both the phases differ from each other. It swarms and migrates in large number.

Economic importance

1. It is very destructive to crops and cause serious losses to farmers.
2. They were used as food by the Matives of Africa and Egypt.

Biological Status

Locust is of international concern. It migrates through African deserts to Arabia, Baluchistan, Pakistan and then to north west India. Locust warning organisation keeps a close vigil on its incidence and occurrence in India and monitors its all activities. Locust hordes create havoc by devastating the crops.

Three species viz. *Cyrtacanthacris* (= Patanga) *succincta* (Bombay locust) *Locusta migratoria* (Migratory locust) and *Schistocerca gregaria* (Desert locust) are commonly found in India.

Painted Grasshoppers (Aak Grass hopper) (*Poecilocerus pictus*) Fig. 3.8 b

Order - Orthoptera

Suborder - Caelifera

Superfamily - Acridoidea

Family - Acrididae

Genus - *Poecilocerus*

Characters of the genus

1. Pronotum is collar shaped.
2. Hind leg is strongly build for jumping during flight.
3. Mouthparts chewing type.
4. Eyes and antennae well developed.
5. Anterior wings are leathery called *tegmina.*
6. Development *hemimetabolous* type.

Habit and Habitat

They feed upon green plants especially on *Calotropis*, otherwise also it prefers many garden plants and milkweeds. It is active during day and night.

Biological importance

It is used in toxicological researches in the field of Entomology.

Economic importance

They damage crops in young and adult stages.

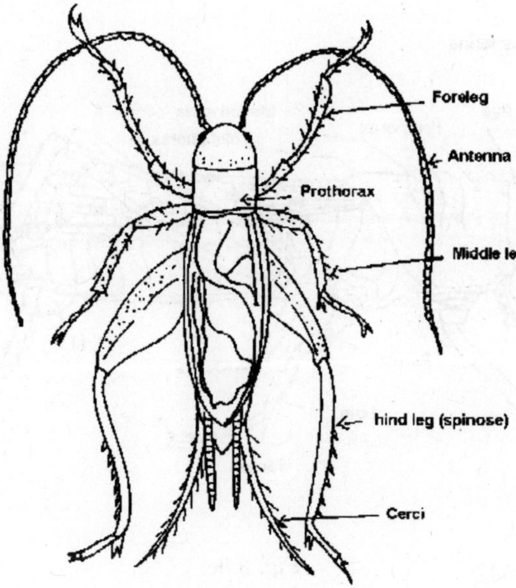

Fig. 3.8a. House cricket

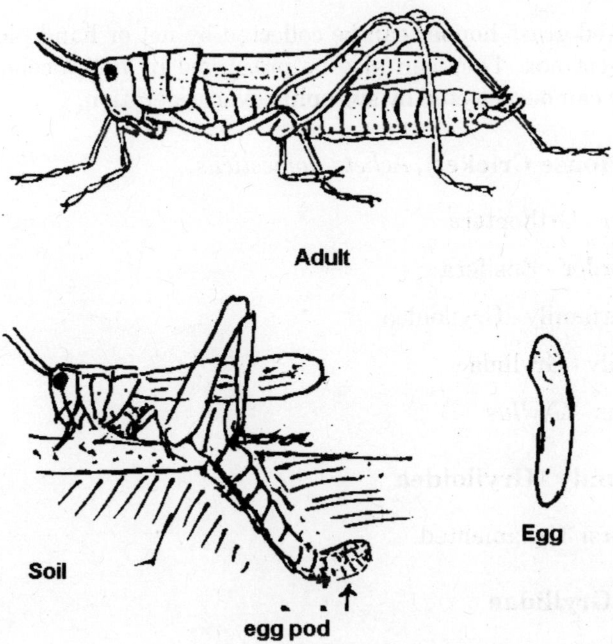

Fig. 3.8b.

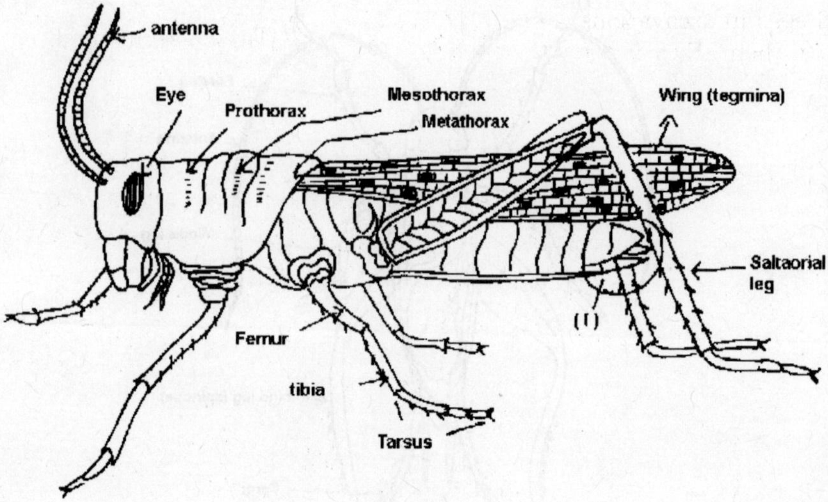

Fig. 3.8c.

Collection

Painted grass hopper can be collected by net or hand picking. Pin in the insect box. They can also be preserved in 70% alcohol in glass jars. They can be collected by sweeping over vegetation.

Gryllus **(House Cricket),** *Acheta domesticus*

Order - Orthoptera

Suborder - Ensifera

Superfamily - Grylloidea

Family - Gryllidae

Genus - *Gryllus*

Super family - Grylloidea

1. Tarsi 3- segmented.

Family - Gryllidae

1. Ovipositor elongated, needle shaped.

Characters of the genus

1. They are *brachypterous* or *apterous.*
2. Prominent eyes and long antennae present.
3. Hind legs strong and helps in jumping.
4. Fore wings are leathery called, *tegmina.*
5. Ovipositor elongated.
6. Mouthparts mandibulate type.
7. Development hemimetabolous.

Habit and Habitat

They are nocturnal and are household pests. They are known to everybody for musical, monotonous chirpings. They are both herbivorous and carnivorous. They are reported cannibalistic in hunger and captivity.

Economic importance

1. Both nymphs and adults damage clothes, starchy material and pollute the food stuffs in the kitchens.
2. The Chinese and Japanese cage the singing crickets and organise `cricket fights'.
3. They sometimes chew the hairs and skin of babies and infants.

Collection

Can be collected manually and pinned. They can also be preserved in 70% alcohol.

Mole Cricket

Family - Gryllotalpidae

1. Fore legs modified for *fossorial life* (They are adapted for digging).
2. Ovipositor small.

Genus - *Gryllotalpa*

Fore legs are adapted for digging i.e. *fossorial*. They are greatly expanded and armed with strong teeth. Tibia expanded and digitate.

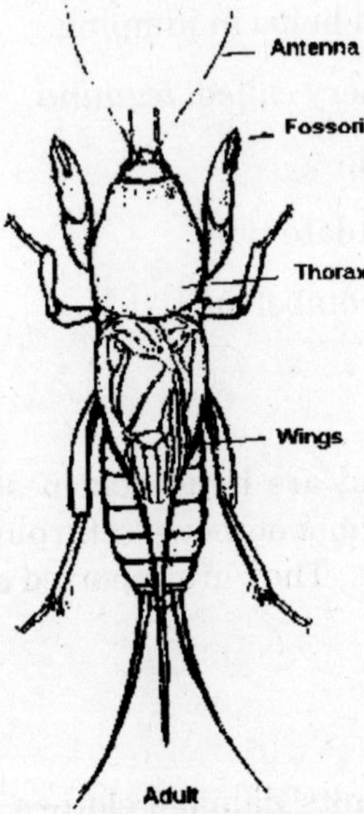

Fig. 3.9.

Habit and Habitat

It is commonly called *mole cricket* because of its habit of burrowing in the soil. Body adapted for sub-terranean life. It eats other insects, worms and plant material. It is seen to be active at night. The eggs are laid in burrows and the life cycle takes about two years. It is also capable of flight.

Collection

Collected manually and pinned. It can also be preserved in alcohol.

Schizodactylus

Order - Orthoptera

Suborder - Ensifera

Super family - Tettigonoidea - (Tarsi 4 segmented).

Family - Schizodactylidae

Genus - *Schizodactylus monstrosus.*

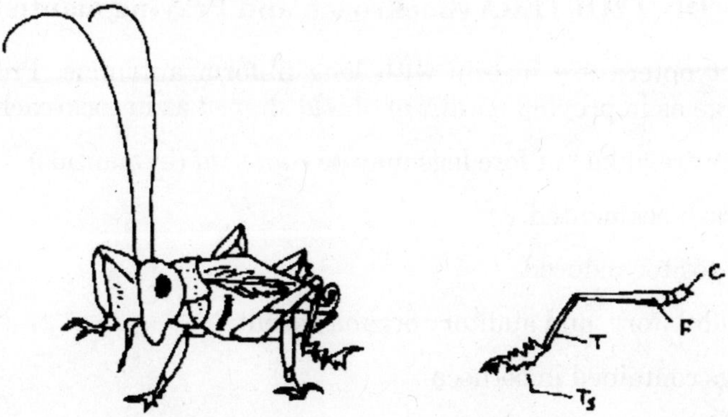

Fig. 3.10. *Schizodactylus* Leg (C - Coxa, F - Femur, T - Tibia, Ts - Tarsus)

Family-Schizodactylidae

1. 2nd and 3rd tarsal segments have minute curious flat expansions.
2. Wings when present coiled spirally at apices during resting stage.

Genus - Schizodactylus

1. It is commonly called *Bheriva* in Bihar.

Habit and Habitat

1. It is burrowing insect living in the sandy soil often near rivers.
2. It is carnivorous but also destructs crops when it burrows and cut the roots.
3. Distribution is dependent on the peculiar conditions of moisture and soil.
4. It is nocturnal and robustly built.
5. The appearance is extremely striking and its large jaws make it appear ferrocious.

Collection

Hand collected and pinned in the box. It can also be preserved in alcohol.

ORDER- DICTYOPTERA (Cockroach and Praying mantis)

1. Dictyoptera are insects with long filiform antennae. Pronotum large as in preying mantis or shield shaped as in cockroaches.
2. Legs are alike but fore legs may be *raptorial* (in mantids)
3. Tarsi 5 segmented.
4. Ovipositor reduced.
5. Stridulatory and auditory organs absent.
6. Eggs contained in ootheca.

It includes two Sub-orders.

1. *Blattoidea* (Cockroaches)
2. *Mantoidea* (Praying mantis)

SUB-ORDER-BLATTOIDEA (= Blattaria, Blattodea)

Some authors have designated *Blattaria* as a separate order, others have placed it as sub order in Dictyoptera. There is considerable confusion concerning the correct designation of this order. Many entomologist place the cockroaches to the family *Blattidae* of order orthoptera. Other names suggested are Blattaria, Blattodea, Blattoidea and Cursoria.

1. Head nearly or completely covered from above by large shield like pronotum.
2. Cerci prominent.
3. Swift runners.

Genus -Periplaneta (Family- Blattidae)

1. Antennae long filiform.
2. Mouthparts are typical mandibulate type.
3. Compound eyes present.

4. Legs slender and similar adapted for fast running.

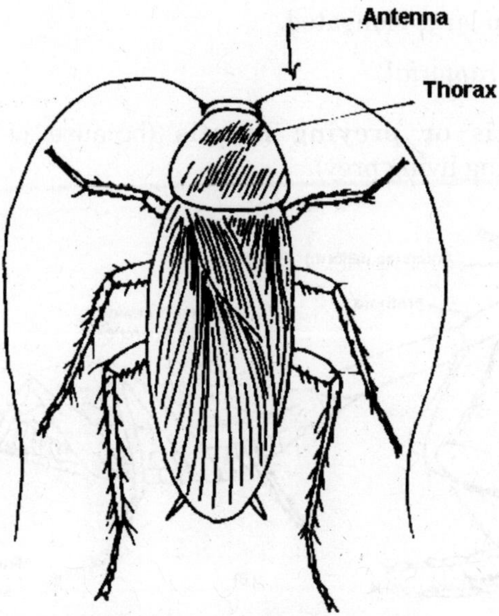

Fig. 3.11. American Cockroach

Habit and Habitat

They are nocturnal but also seen venturing during day time. They feed upon everything whatever they find edible (*omnivorous*).

Economic importance

They are household pests and spoils kitchen articles and food. They cause nuisance to human beings.

Biological importance

They are used in toxicological researches in the field of Entomology.

Collection

Hand collected and pinned in the box. They can also be preserved in alcohol.

SUB-ORDER - MANTOIDEA

1. Pronotum large elongated.
2. Forelegs *raptorial.*

Praying mantis or preying mantis (because of their peculiar attitude in seeking living prey).

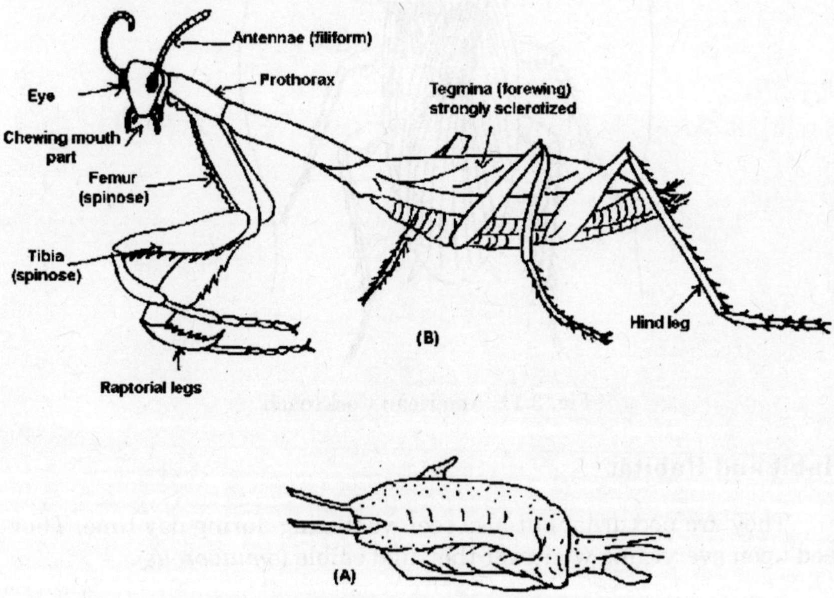

Fig. 3.12. (A) Ootheca, (B) Praying mantis

Family -Mantidae

1. Fore legs are *raptorial* i.e. adapted for ruptering the body of the prey. They are raised together infront and helps in seizing the prey.
2. Antennae long.
3. Head, small, triangular with vertical face and movable on slender neck.
4. Eggs are deposited in ootheca or capsule formed of frothy quick drying material.
5. Style on 9th segment and cerci on 10th segment.

Habit and Habitat

1. They are largely arboreal and occurs in all warmer parts of the world. Many of the mantids resembles immediate environmental conditions as a result their victims are deceived.
2. They are carnivorous and leads predaceous life both as young and adult.
3. They feed on flies, grasshoppers, crickets, caterpillars etc.
4. They deposit their eggs in ootheca in two or more rows. The nymphs are prone to *cannibalism.*
5. The females usually devours the males after mating.

The males are caged and used for competitive mantid fights in China.

Collection

Hand collected and pinned in the box. They can also be preserved in alcohol.

ORDER-PHASMIDA (Stick insects and leaf insects)

1. Prothorax is very short, meso and meta-thorax are usually elongated.
2. Head *prognathous* and mouth- parts mandibulate.

Characters of Stick insect

1. Large, apterous elongated insects.
2. Tarsi 5 segmented.
3. Antennae long (8-10 segmented)
4. Abdomen 11 segmented.
5. Eggs are laid singly.
6. *Autotomy* is found in nymphal legs. The legs can be detached between trochanter and femur.

Habit and Habitat

1. They are arboreal, nocturnal, solitary and sluggish. All are phytophagous.

2. They are predominantly tropical insects remarkable for their close protective resemblance to foliage (*mimicry*) twigs or vegetation.

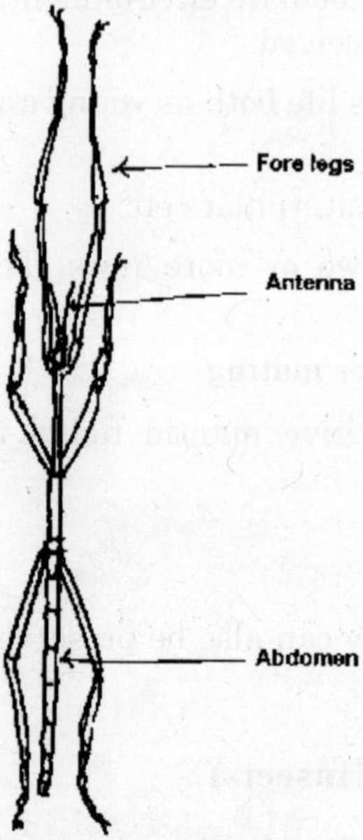

Fig. 3.13a. Stick insect

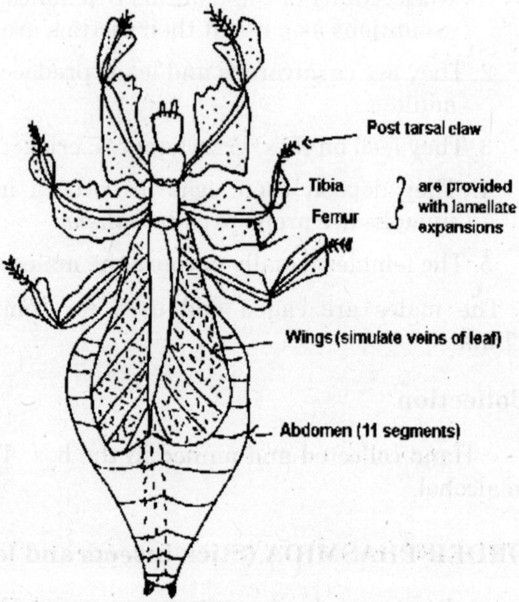

Fig. 3.13b. *Phyllium*

Economic importance

Sometimes they occur in large number in forests and defoliate the trees.

Biological peculiarities

1. They, exhibit protective colouration (*mimicry*).
2. *Autotomy* is found in nymphal legs. *Regeneration* of the lost part takes places.
3. Parthenogenesis.
4. They can change colour.
5. They lay eggs in ootheca.
6. Death feigning present.

Collection

Hand collected and pinned in the box. They can also be preserved in alcohol.

Phyllium (Leaf insect or Walking leaves)

1. Body broad and flat.
2. They simulate the surroundings. Not only they simulate the colours and shapes of leaves but even they show the veining as well and the legs have leaf like appendages.
3. They are the inhabitants of the moist tropical regions.
4. Specimens are comparatively rare in most insect collections.
5. Tegmina greatly enlarged. The front costal portion of the wings is often thickened and pigmented like the tegmina while the membranous areas are transparent or brilliantly coloured. The tegmina and wings are folded flat on the back at rest.

Collection

Hand collected and pinned in the box. Can also be preserved in alcohol. They can also be picked with the help of forceps.

ORDER - DERMAPTERA (Earwigs)

1. Fore wings truncated, leathery whereas the hind wings are semi circular, membranous and greatly extended in anal area. Pre-anal part of hind wing is mainly sclerotized. The veins are radially arranged.
2. Cerci are unjointed and always modified into heavily sclerotized forceps. The presence of forceps is a distinctive feature of the order.
3. Metamorphosis incomplete.

Characters of Earwigs (*Forficula*)

1. Antennae many segmented.

2. Mouthparts - *chewing* type.

3. Tarsi 3 segmented.

4. Abdomen 11 segmented.

5. Exhibits *parental care*.

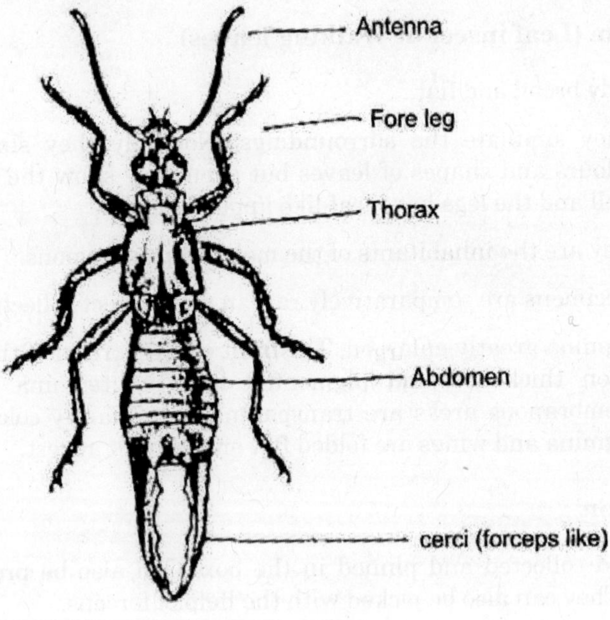

Fig. 3.14. Earwig

Habit and Habitat

1. They are nocturnal and hide during the day under stones, rocks, leaves and bark of trees, small crevices etc.

2. They are attracted to lights at night.

3. They are primarily *carnivorous* but are probably *omnivorous*.

In Mediterranean region it is predaceous on maggots and scavanger. It is noxious as pest in household where its noctunal habits are similar to cockroaches.

Biological peculiarity

The unique character of earwigs is the display of parental solicitude. After fertilization the female lays 40-50 eggs in excavations in the ground. Then it sits over the eggs and guard them. Even after the hatching the young ones are protected by the mother. The life cycle is completed in about six months.

1. Collection

Hand collected and pinned in the box. They can be preserved in alcohol also.

Note : There is a strong belief that the earwigs enter the ear of a sleeping man. But there is no evidence for this accusation.

ORDER - EMBIOPTERA (Web spinners)

1. Tarsi 3 segmented. First tarsal segment of the fore legs is swollen in both the sex, which contain glands and spinnerettes.
2. They spin silken tunnels.
3. Mouthparts *chewing* and *biting* type.
4. Female always apterous.

Web spinners

1. Exhibit *sexual dimorphism.* Females are apterous. 1. Cerci segmented 2. Wings smoky similar with micro trichina.

Habit and Habitat

1. They are nocturnal.
2. They are gregarious and lives in silken tunnel.
3. They occur in humid places. Larvae feeds on decaying vegetable matter. Males are predatory.

Biological peculiarity

They prepare silken lined tunnels as the fore metatarsus is swollen and contain the spinning glands.

Collection

Hand collected and pinned. Can also be preserved in alcohol. Slide can also be prepared. Locate the silken tunnel and disturb it collect the insect by cotton plug doused in 70% alcohol.

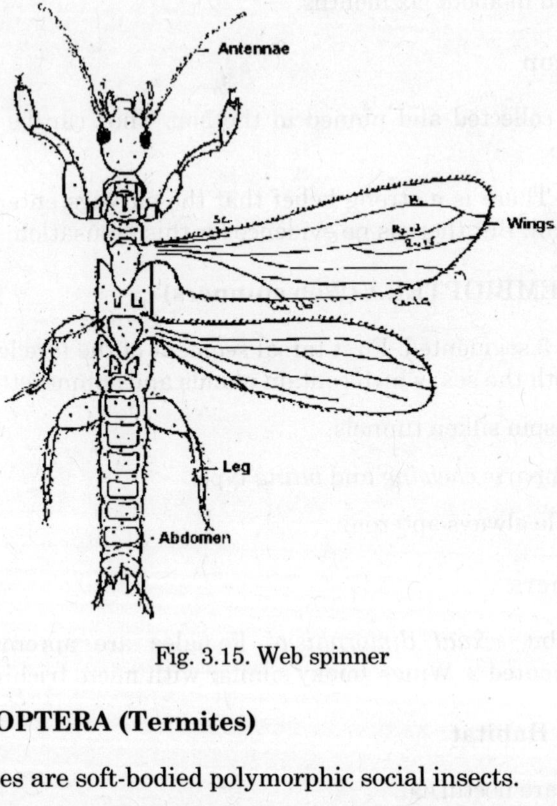

Fig. 3.15. Web spinner

ORDER-ISOPTERA (Termites)

1. Termites are soft-bodied polymorphic social insects.
2. Wings two pair present in sexually mature males and females, similar, deciduous. They are shed by means of basal fractures. Anterior veins are markedly sclerotized, cross veins few or absent.
3. Mouthparts *biting* and *chewing* type.
4. Antennae short *moniliform*.
5. Cerci present.
6. Metamorphosis slight.

Habit and Habitat

1. Termites live in big nest called *termitaria* which are sub terranean and which sometimes rise above the soil surface as large mounds. *Termitaria* consist of tunnels and chambers.
2. They feed on wood and are able to digest cellulose with the help of protozoa in their gut. They exhibit symbiosis with intestinal protozoans. They also grow fungal gardens to satisfy their nutritional requirements.
3. They avoid light and move about along covered runways.

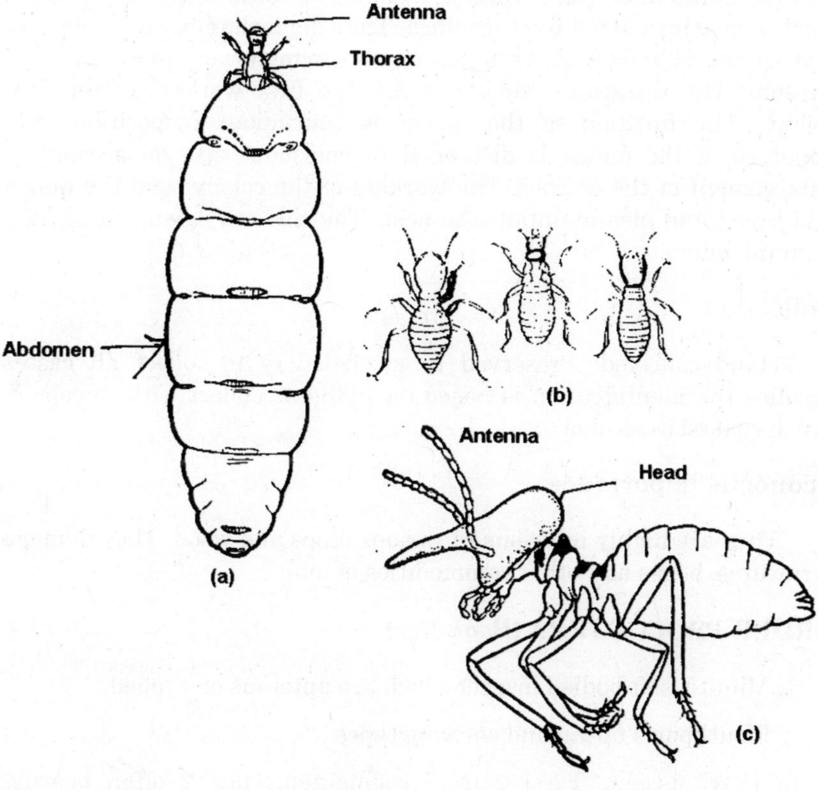

Fig. 3.16. (a) Queen termite, (b) Workers, (c) Nasute soldier.

Biological peculiarities

1. *Symbiotic flagellates* (Polymastigina and Hypermastigina) assist digestion of cellulose in most termites.

2. *Neoteny* is wide spread among termites and occurs in diverse forms such as macropterous sexual adult, brachypterous sexual neotenic.
3. They are most prolific of all insects.
4. The *reproductive castes* are sexually mature males and females and comprise the king and queen. The *Sterile castes* includes the workers and soldiers (apterous).

The reproductive castes leaves the nest and fly in swarms during evening hours after rains. After a while, they settle down in pairs and shed their wings along fracture lines. They dig a chamber and eggs are laid in the chamber. After hatching the nymphs are tended by the parents. The resulting individuals are the first workers of the new colony. The function of the queen is continuous oviposition. The abdomen of the queen is distended to enormous size on account of enlargement of the ovaries. The workers of the colony tend the queen and brood and also maintain the nest. The soldiers defend the colony from intruders.

Collection

Hand collected. Preserved in alcohol. Try to collect all castes because the identification is based on soldiers. Collect with forceps & brush soaked in alcohol.

Economic importance

They are highly injurious to various crops and wood. They damage furnitures, books and other commodities of man.

ORDER-PSOCOPTERA (*Book lice*)

1. Minute, soft bodied insects which are apterous or winged.
2. Mouthparts *biting* and *chewing* types.
3. Cerci absent. Tarsi 2 or 3 segmented, claw 2 often bearing suctorial setae.
4. Antennae setaceous with 13 or more segments.

Habit and Habitat

1. Psocids often live gregariously on plants, under bark, fungi, lichens, algae fallen leaves, in decaying vegetable matter. They are found in damp warm places in houses and store rooms.
2. Some species are predaceous on eggs of coccids or on aphids.
3. A number of species occur among old papers and books, preserved museum specimens, stored tea, cereals like rice etc.

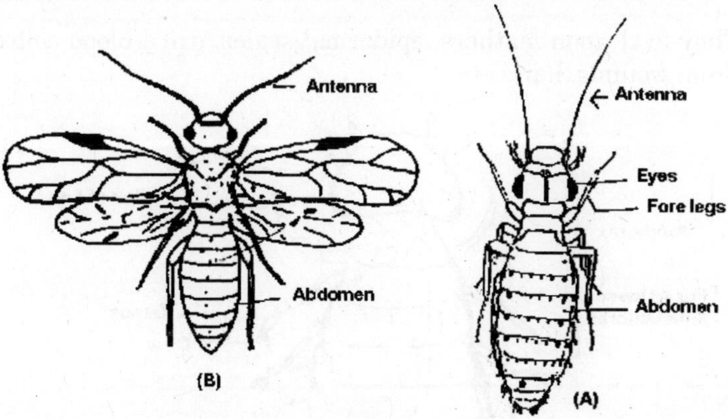

Fig. 3.17. Booklice (a) apterous, (b) winged.

Economic importance

They are injurious to paper and books.

Collection

Hand collected with the help of camel hair brush soaked in alcohol. Prepare a permanent slide. Book lice which cause extensive damage to documents, specimens etc. can be collected by jerking them on enamel tray and picking up by brush.

ORDER - MALLOPHAGA (Biting lice or Bird louse)

1. Wingless ectoparasite of birds.
2. Head broader than or same as abdomen, meso and meta-thorax separated from prothorax.
3. Mouthparts *biting* type.
4. Antennae short incospicuous. Eyes reduced.

5. Cerci absent
6. Legs are adapted for *clinging.*

Habit and Habitat

1. Commonly called bird lice. They are entirely ectoparasitic on warm blooded animals specially bird.
2. They feed upon feathers, epidermal scales, dried blood collecting from wounds, hairs etc.

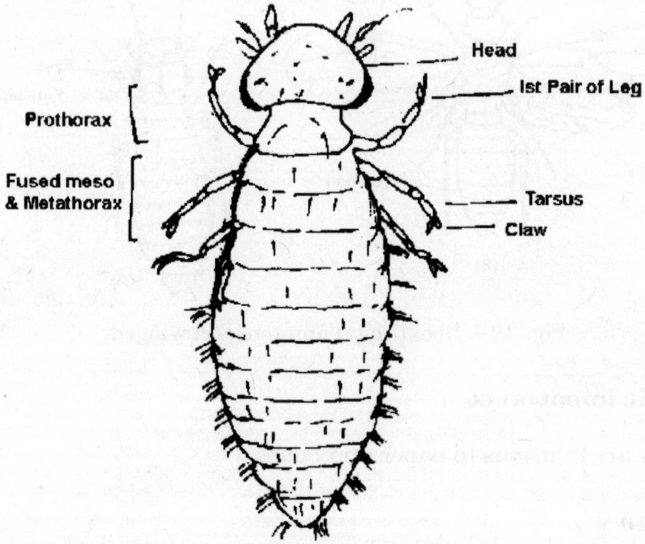

Fig. 3.18. Bitting Lice (Mallophaga)

Parasitic adaptations

Body size small, stream lined body, dorso ventrally flattened, body covered with bristles, wings absent, integument hard and elastic providing hydrostatic protection for internal organs, claws well developed for grasping feathers and hairs.

Economic importance

1. They cause intense irritation to the host and when present in large numbers often leads to its death by nervous break down.

2. They are injurious to poultry (genus -Menopan) genus Lipeurus on domestic fowl.
3. Some species also cause loss to wool and live stock.

Collection

From the infected bird body with the help of brush soaked in alcohol. Prepare a permanent slide and preserve in 70% alcohol.

ORDER - ANOPLURA OR SIPHUNCULATA (Sucking lice or Human louse)

1. Completely wingless (secondarily wingless) insects living as ectoparasite of mammals.
2. Mouthparts are *piercing* and *sucking* type and *entognathous*.
3. Claws specialized for grasping hairs.
4. Antennae short (at the most 5 segmented).

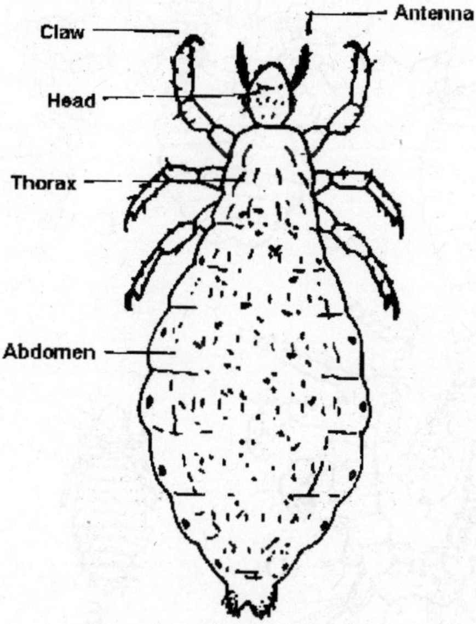

Fig. 3.19. Louse (Anoplura)

Habit and Habitat

1. They are ectoparasite of man and mammals.

2. The eggs are laid in clothing or attached to hairs.

3. They suck blood of man or other mammals.

Economic importance

Pediculus transmits dreaded typhus fever in man. Other diseases transmitted by the human louse are **trench fever** and **relapsing fever.**

Collection

Hand collected. Prepare a permanent slide and preserve in alcohol also. **Head louse** - *Pediculus humanus capitis;* **Body louse** - *Pediculus humanus humanus;* **Crab-louse** - *Phthirus pubis* collect- from pubic hairs and eye lashes in the patients of Blepharitis. **Buffalo louse** - *Haematopinus tuberculatus,* **Horse louse** - *Haematopinus usini,* Pig louse - *Haematopinus suis.*

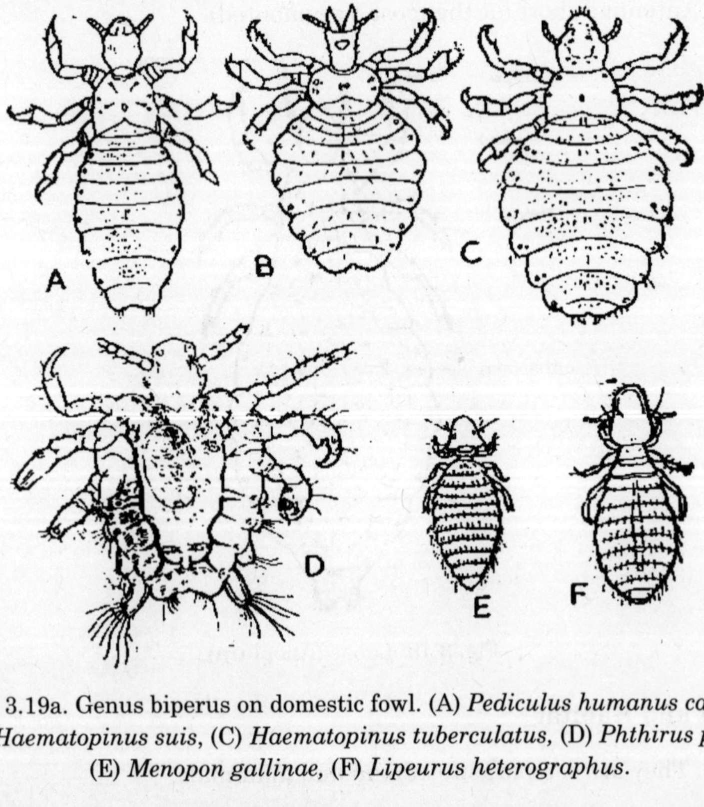

Fig. 3.19a. Genus biperus on domestic fowl. (A) *Pediculus humanus capitis,* (B) *Haematopinus suis,* (C) *Haematopinus tuberculatus,* (D) *Phthirus pubis,* (E) *Menopon gallinae,* (F) *Lipeurus heterographus.*

ORDER - HEMIPTERA (Bugs)

1. Fore wings are not uniform in texture i.e. harder at the base and membranous at the tip (*hemelytron*).
2. Mouth parts *piercing* and *sucking* type. Needle like styles which are the modified mandibles and maxillae, are enclosed in a groove on the dorsal surface of the labium which forms the *rostrum*.
3. Metamorphosis incomplete.

Hemiptera is divided into two sub-orders, (i) *Heteroptera* and (ii) *Homoptera*.

The bugs with *hemelytron* are placed under Hetroptera. The insects belongings to Homoptera possess fore wings of uniform thickness. Also the wings of Homoptera fold roof like over the body at rest.

Series *Gymnocerata* includes terrestrial insects which are phytophagous generally. Some species however, are predaceous and blood sucking. **Series** *Cryptocerata* includes all aquatic bugs which are largely predoceous in habit. *Auchenorhyncha* have very short antennae with a terminal arista. The rostrum arise plainly from the head. Tarsi 3 segmented and are active insects. *Sternorhyncha* do not show prominent terminal arista on the antennae. Rostrum arise between the anterior coxa. Tarsi are 1 to 2 - segmented.

Bugs are injurious to crops as they suck the sap of plants.

ORDER - HEMIPTERA

Sub-order

- Heteroptera
  - Series
    - Gymnocerata
      - Families
        1. Reduviidae
        2. Gerridae
        3. Cimicidae
        4. Lygaeidae
        5. Pyrrhocoridae
        6. Coreidae
        7. Tingidae
        8. Pentatomidae
        9. Miridae
    - Cryptocerata
      - Families
        1. Belostomidae
        2. Nepidae
        3. Notonectidae
        4. Corixidae
- Homoptera
  - Auchenorhyncha
    - Families
      1. Cicadidae
      2. Cercopidae
      3. Jassidae
      4. Membracidae
      5. Fulgoridae
  - Sternorhyncha
    - Families
      1. Coccidae
      2. Aleurodidae
      3. Aphidae

Reduviid Bugs (Assasin bugs or Kissing bugs)

Hemiptera

Hetroptera

Gymnocerata

Reduviidae

1. Proboscis attached to the front end of the head and is curved.
2. Fore legs *raptorial*.
3. Antennae *filiform*.

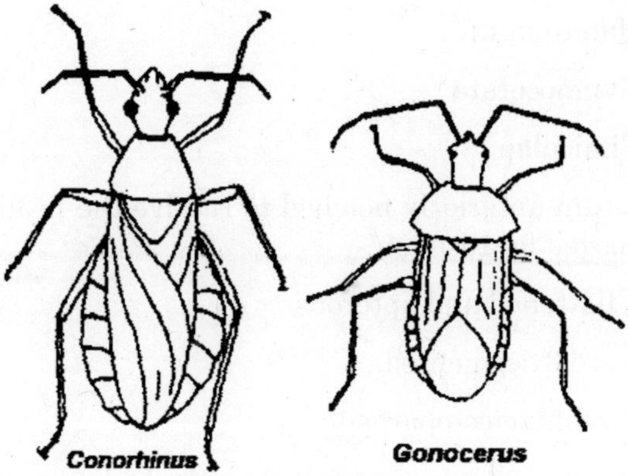

Fig. 3.20.

Habit and Habitat

1. It is predatory on other insects which frequents fruit trees. They prey upon various soft bodied caterpillars. Some are blood sucking ectoparasites like Rhodnius. *Rhodnius* does not occur in India.
2. *Conorhinus* (assasin bug) sometimes attacks man. It is often found in household.

Economic importance

1. On account of its blood sucking habit it is reported to carry disease causing micro organisms.
2. They kill many injurious insects.

Collection

Hand collected and pinned in the box. Can also be preserved in alcohol. They can also be collected by sweeping net. Beating branches with a stick over a sheet of cloth.

Note : The name `kissing bugs' arise from a news in 1899 when one of these bugs pierced the lip of a lady.

Cimex (Bed bug)

Hemiptera

Heteroptera

Gymnocerata

Cimicidae

1. Pronotum anteriorly notched to receive the head and ventrally it is grooved for rostrum.
2. Oval, flattened and apterous.
3. Tarsi are 3 segmented.
4. Blood sucking ectoparasite.
5. *Odoriferous glands* are present on the body.

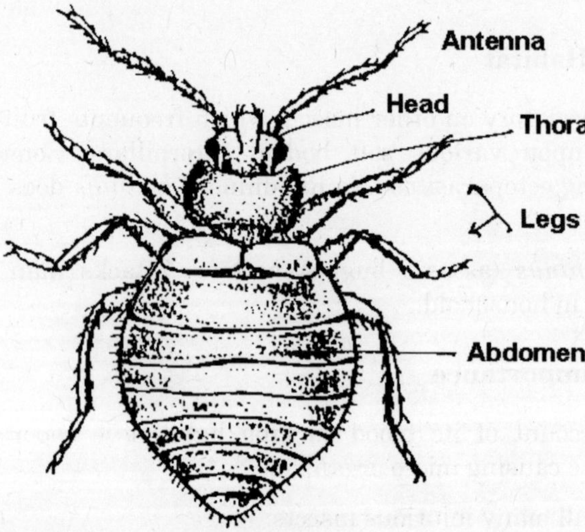

Fig. 3.21. *Cimex*

Habit and Habitat

1. They are blood sucking ectoparasites on warm blooded animals.
2. They live concealed in the furniture of cinema house, buses, railway trains and cots.
3. When handled, they emits a repulsive odour.

Economic importance

1. The bite of the bed bugs is very annoying and painful due to itching sensations. They transmitts many diseases. They suck the blood of man.

Collection

Hand collected Prepare a permanent slide. Use forceps to collect.

Pond skaters or Water striders

Hemiptera

Heteroptera

Gymnocerata

Gerridae

Gerris

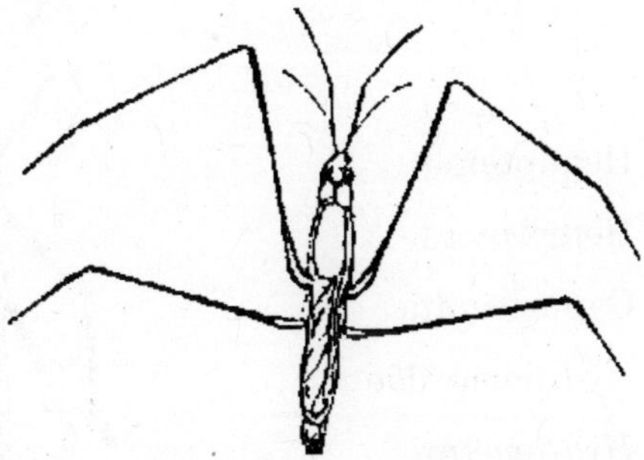

Fig. 3.22a. Water strider

1. Thorax is the widest part of the body and as long as abdomen.
2. Hind femora very long extending beyond the apex of the abdomen.
3. Middle and Hind legs arise close to each other.
4. Front legs are placed much forward.
5. Front legs are more or less *raptorial*, only the middle and hind legs are used for locomotion.

Habit and Habitat

1. They are found running over the surface of water. Their locomotion looks like skating hence the name *Pond skaters*.
2. They feed on dead insects floating on the water surface caught by the front legs.

Collection

With the help of nylon net. Pin in the insect box. Can also be preserved in alcohol.

Halobates frequents the surface of the sea.

Hydrometra

Hemiptera

Heteroptera

Gymnocerata

Hydrometridae

Hydrometra

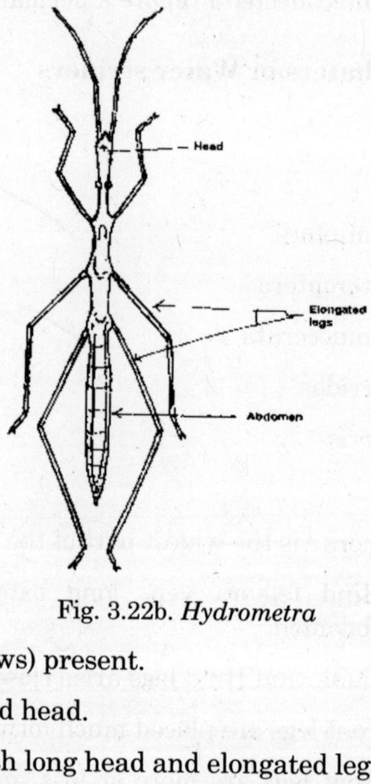

Fig. 3.22b. *Hydrometra*

1. *Pulvilli* (pad below the claws) present.
2. Ocelli absent, apical dilated head.
3. Delicate rod like insect with long head and elongated legs.

Habit and Habitat

1. It is found crawling slowly on the surface of the water.
2. It feeds upon small insects.

Collection

With the help of nylon net. Prepare a permanent slide and preserve in the alcohol.

Dysedercus (Red cotton bug)

Hemiptera

Heteroptera

Gymnocerata

Pyrrhocoridae

Dysdercus.

1. It is bright red in colour with black spots on the wings.

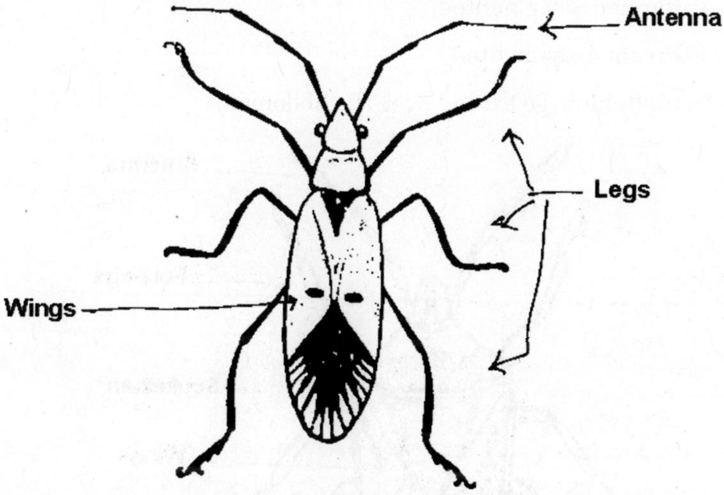

Fig. 3.23. Red Cotton bug

Economic importance

1. It is serious pest of cotton and other plants of *Malvaceae* like Bhindi (Okra).
2. It feeds on cotton bolls and stains the lint and thus cause severe loss to farmers.

Collection

Hand collected and are pinned in the box.

Nezara

Hemiptera

Heteroptera

Gymnocerata

Pentatomidae

Nezara

Nezara viridula is a greenish bug.

1. Pentatomids are commonly called *stink bugs*.
2. Antannae 5-segmented.
3. Rostrum 4-segmented.
4. Scutellum large extend over the abdomen.

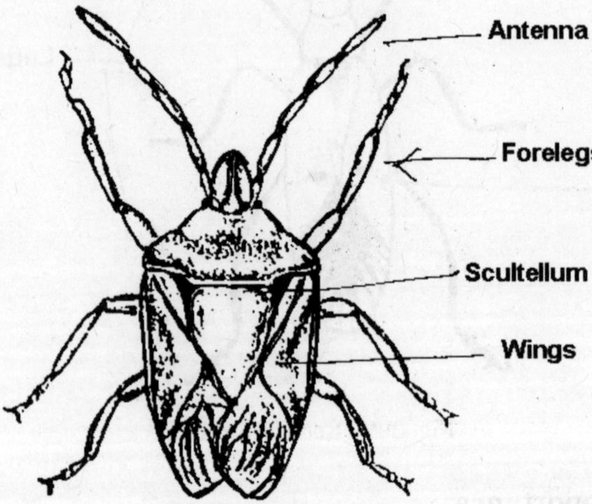

Fig. 3.24. *Nezara*

Habit and Habitat

1. It is a pest of vegetable crops and sun hemp.
2. The bug produces obnoxious smell from its *odoriferous glands*.

Economic importance

It is pest of vegetable crops like brinzal etc., sun hemp and other useful crops.

Collection

Hand collected or with the help of net. Pin in the box.

Belostoma (Giant water bug)

Hemiptera

Heteroptera

Cryptocerata (Antennae hidden under side of the head).

Belostomidae

Belostoma

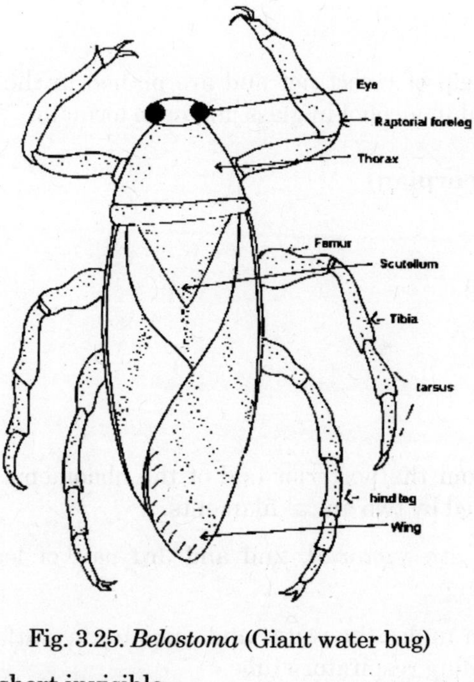

Fig. 3.25. *Belostoma* (Giant water bug)

1. Antennae short invisible.
2. Legs adapted for swimming (*natatorial*) and are with complex respiratory modifications.
3. Fore legs are *raptorial*.
4. The concave abdominal dorsum underwings forms a air reservoir.
5. Body large, flat and broad.
6. Rostrum, short, curved and segmented.

Habit and Habitat

1. They are aquatic and usually attracted to light.
2. They feed upon insects, fish, snails tadpoles and froglets or toad lets.

Economic importance

Sometimes they bite the swimmers. Their claws are sharp and able to cause painful punctures.

Collection

With the help of insect net and are pinned in the box. They can also be preserved in alcohol in glass jars or in formalin.

Nepa (Water scorpian)

Hemiptera

Heteroptera

Cryptocerata

Nepidae

Nepa

1. Arising from the posterior end of the abdomen is a respiratory tube formed by two apical filaments.
2. Fore legs are *raptorial*, 2nd and 3rd pair of legs adapted for walking.
3. They swim under the water and come to the surface to take air by protruding respiratory tube.
4. Lay their eggs in aquatic plant tissues.

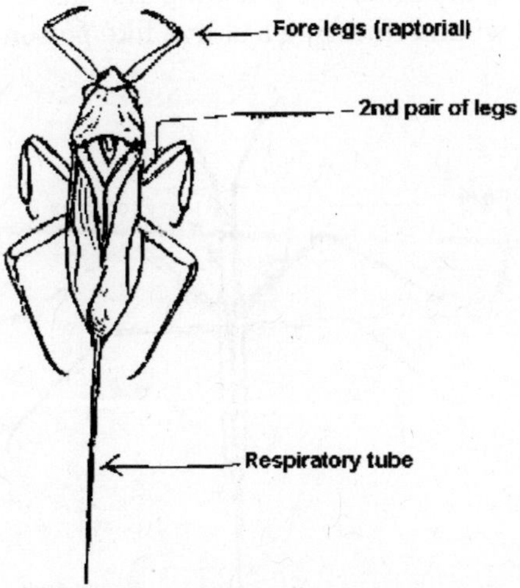

Fig. 3.26. *Nepa*

Habit and Habitat

1. Commonly called *Water scorpian* because of their general resemblance to a scorpian.
2. Commonly found in fresh water and often come to light.
3. They feed on other aquatic insects. They are *predaceous*.

Collection

With the help of net. Pin in the box or can also be preserved in alcohol or formalin in glass jars.

Ranatra

Classification same as Nepa.

1. Fore legs strongly *prehensile* and mid and hind legs are with spines.
2. Abdomen with a apical breathing tube which is made up of two filaments.
3. Female with well developed pointed toothed ovipositor.
4. Long prothorax forms the neck.
5. Coxae are much developed and long.

6. Fore legs are modified for grasping the prey while the piercing beak inflict wound injecting a saliva like poison.

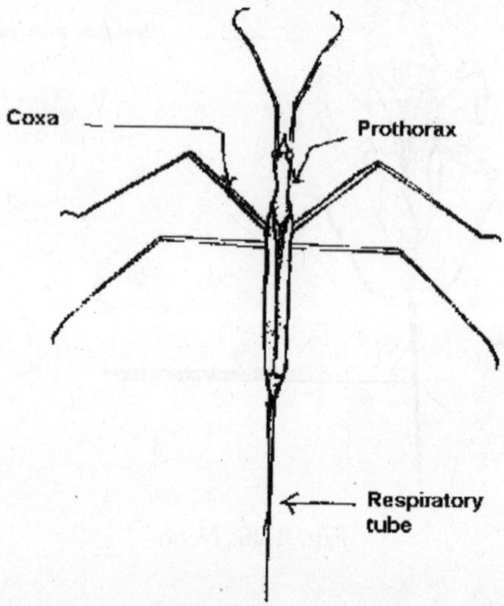

Fig. 3.27. Ranatra

Photo. Natural habitat of Nepa, Ranatra and Back swimmers.

Habit and Habitat

They are found in fresh waters and are *carnivorous*.

Collection

With the help of nylon net. Pin in the box, can also be preserved in alcohol.

Notonecta (Back swimmer)

Hemiptera

Heteroptera

Cryptocerata

Notonectidae

Notonecta

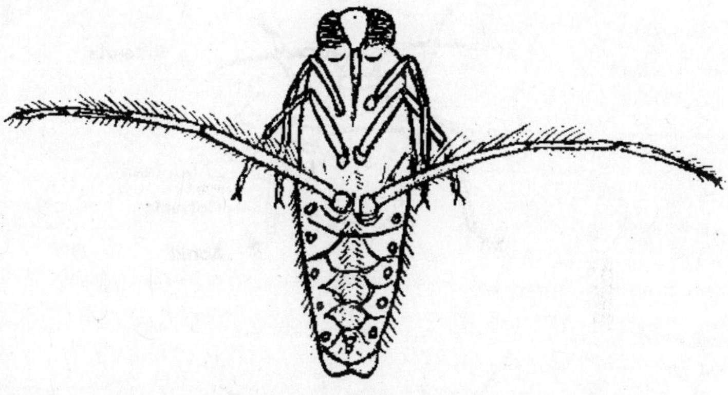

Fig. 3.28. *Notonecta*

1. Body dorsally convex.
2. Head is overlapped by prothorax.
3. Respire by means of air stores under the wings and they are capable of flight.

Habit and Habitat

1. Commonly called *Back swimmers* as they lie on their back while swimming.
2. They are found in fresh water lakes and ponds.
3. They feed on frog tadpoles, small fish and other insects.

Collection

With the help of nylon net and all pinned in the box. Can also be preserved in alcohol.

Aphid (Plant lice)

Hemiptera

Homoptera

Sternorhyncha

Aphididae

Aphid

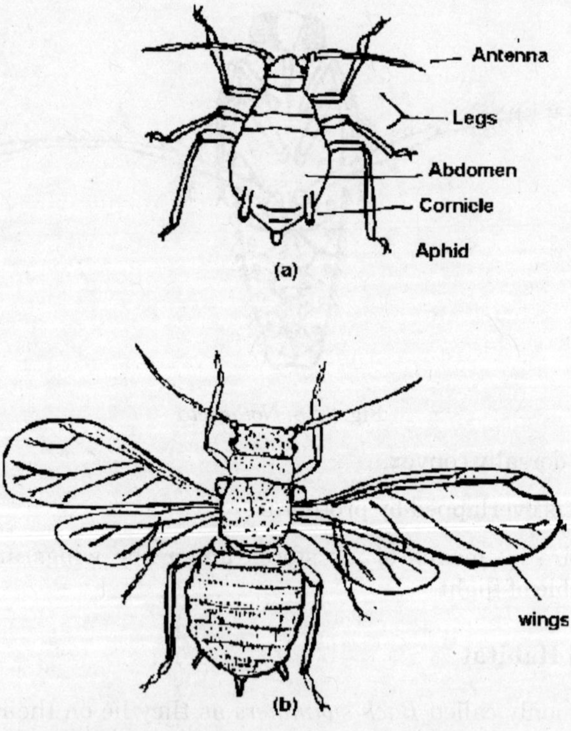

Fig. 3.29. Aphids (a) apterous, (b) winged.

1. Pear shaped small soft bodied insect.
2. Antennae and legs are long.

3. Both winged and wingless forms present.
4. The wings are delicate and show a few simple or branched veins.
5. Rostrum is 4 segmented.
6. A pair of tubes called *Cornicles* are present on the back of the sixth abdominal segment through which they secrete waxy substance.

Habit and Habitat

1. They suck the sap of the plants hence commonly called *plant lice.* They are *polyphagous.*
2. Appears in plains in cold season.
3. Migration takes place.

Economic importance

1. They attack a wide variety of crops, vegetable crops, garden plants. They are serious pest of many cruciferous plants like mustard. They cause heavy loss to farmers as the plant dies and yield is reduced.
2. They also transmits many plant diseases as they act as vectors for plant viruses.

Biological peculiarity

1. Life cycle is complicated and exhibits polymorphism and alternation of generations among parthenogenetic, oviparous and viviparous generations.
2. Winged females exhibits migration to new food plants.

Collection

Hand collected with camel hair brush soaked in alcohol. Prepare a permanent slide.

Cicada

"Cicada has a fortunate life as it has a noiseless wife".

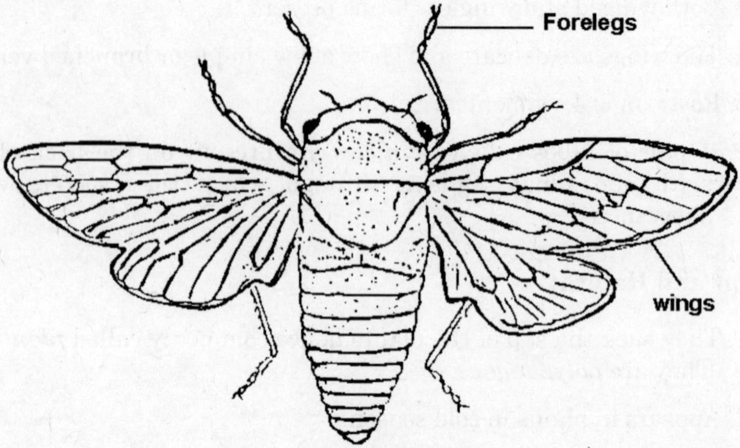

Fig. 3.30. Cicada

This phrase is very famous for Cicada.

Hemiptera

Homoptera

Auchenorhyncha

Cicadidae

Cicada

1. Fore femora greatly swollen, spined beneath and adapted for digging.
2. Tympanal organ lies in the 1st abdominal tergite. Sound is produced by the vibrations of drum.
3. Only the male is capable of producing sound. Female can only hear the sound with the help of auditory organ.

Habit and Habitat

1. The newly hatched nymph dig the soil with the powerful *fossorial leg* and spent long life undergound.
2. They suck the sap of the roots of the plants.
3. In Seventeen year cicada (*Tibicen septemdecim*) the nymph comes from the ground after 17 years in May.

Economic importance

1. Many species of Cicada are used as human food in different parts of the world.
2. Blistering cicada causes blisters on human skin in Japan, China and south-east Asia.

Biological peculiarity

1. Cicadas are known for the sound they produce. Only male can produce the sound which is produced by the vibrations of the drum in which has a resonating air chamber. The sound is deafeningly loud.
2. 17 year Cicada takes 17 years to become adult.

Collection

Can be collected with the help of net and are pinned in the box. It can also be preserved in alcohol or formalin jar.

Photo. Natural habitat of Cicada.

ORDER - THYSANOPTERA (Thrips)

1. Bristle like wings.

2. Mouthparts *piercing* type and asymmetrical i.e. labrum is of irregular form, right mandibles reduced or absent, left mandible is well developed.
3. Tarsi are very short, 1 or 2 segmented. The last tarsal segment has a cup like depression in which there is a bladder like structure.
4. Cerci absent.
5. Antennae 6-10 segmented.
6. Metamorphosis shows characters of holometabola. There is a pupa in the life cycle.

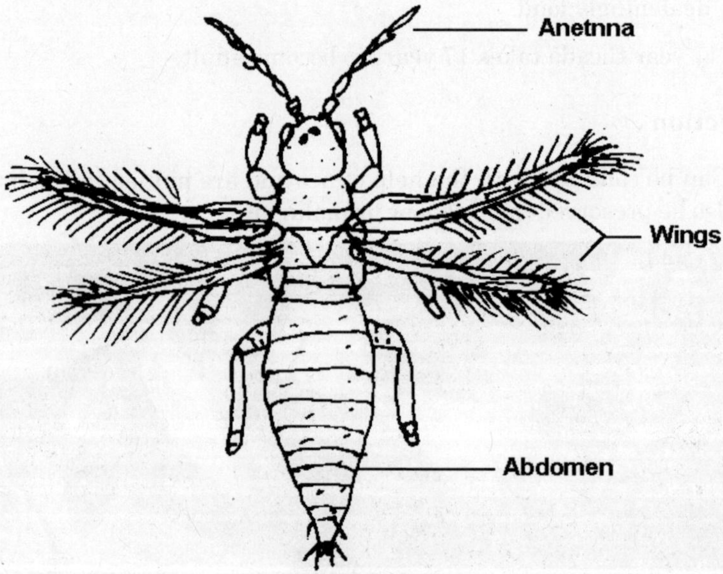

Fig. 3.31. Thrips

Habit and Habitat

1. Found usually on flowers and other parts of the plants. Some live under bark in decaying vegetable matter and some may form galls.
2. They suck liquid food. Most species suck plant sap but some species suck body fluids of the other insects and eggs of mites and lepidoptera.

Economic importance

1. Many species are pest of plants like citrus, cotton, cereals, tomato, tobacco and potatoes. *Thrips tabaci* is the most harmful. It is known as tobacco thrips or onion thrips.
2. They also act as carriers of plant diseases.
3. Some species arc useful as they pollinate flowers and are also used for biological control of weeds. Some species **form galls**.
4. They transmit spotted wilt of tomato.

Biological peculiarity

1. Sometimes they fly in swarms and get into the eyes of human beings and cause annoyance.

Collection

Hand collected. Prepare a permanent slide. Jerk or sweeps the flowers in a alcohol filled petridish and collect the thrips if any.

ORDER - NEUROPTERA (Ant lions)

1. Two pairs of wings are similar and folded roof like over the body when the insect rests.
2. Mouthparts are of *biting* types.
3. Cerci absent.

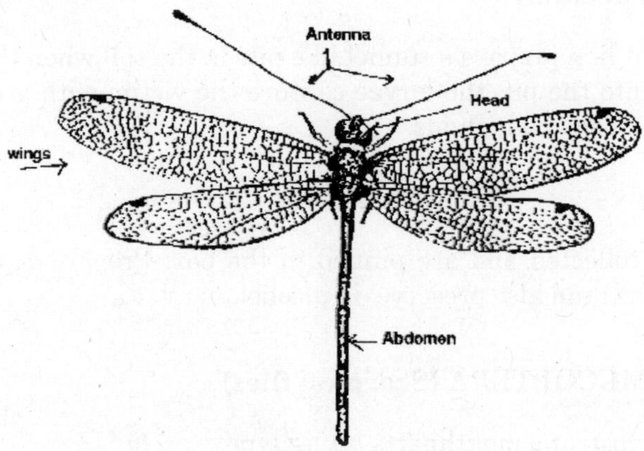

Fig. 3.32. Neuroptera (Lacewings, Ant lion)

Habit and Habitat

1. Both adult and larvae are predaceous on other insects. The larva of *Myrmeleon* is called *ant lion* on account of its feeding on ants.
2. *Ant lions* are often attracted to light at night. During the day it hides amidst the vegetation.
3. *Chrysopa* feeds on aphids and the *Mantispa* on eggs and young ones of spiders.

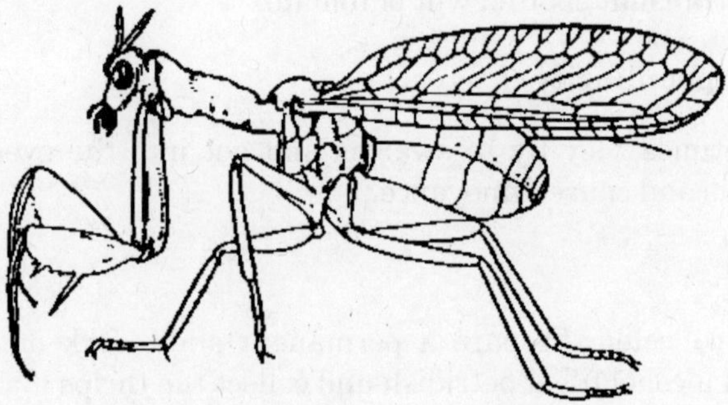

Fig. 3.33. An adult *Mantispa* sp.

Economic importance

Chrysopa feeds on aphids thus act as agent of biological control of aphids. The larvae of Myrmelon feeds on ants.

Biological peculiarity

The ant lion prepare a tunnel like pits in the soil when the unwary ants slips into the pit, the larvae capture the victim with great agility and sucks out the body fluids.

Collection

Hand collected and are pinned in the box. Prepare a permanent slide of larvae and also preserve in alcohol.

ORDER - MECOPTERA (Scorpion flies)

1. Head rostrate, mouthparts *biting* type.

2. Wings absent or vestigial or well developed, membranous, similar, held roof like or flat over the body.

3. Antennae *filiform* with 16-20 or 40-50 segmented.

4. Genitalia of male usually greatly swollen.

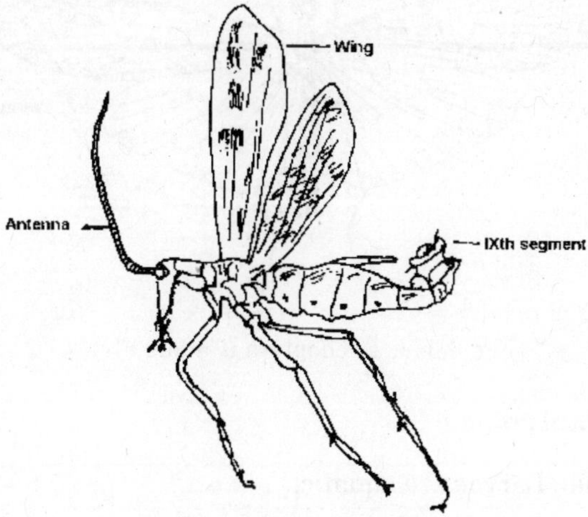

Fig. 3.34. Scorpion Fly.

Habit and Habitat

The larvae are predaceous or often feeds on moss and decaying vegetable matter. Some adult feeds on Nectar and petals of flowers. Pupation takes place in soil, rotten logs and vegetable debris.

Collection

Can be collected by gently sweeping or beating the foliage. Stones should be jerked, wingless forms can be picked up by a brush. dipped in alcohol.

ORDER- TRICHOPTERA (Caddis flies)

1. Moth like

2. Mouthparts *biting* type and vestigial.

3. Wings 2-pairs, membranous (hairy).

4. 1-2 articulated cerci.

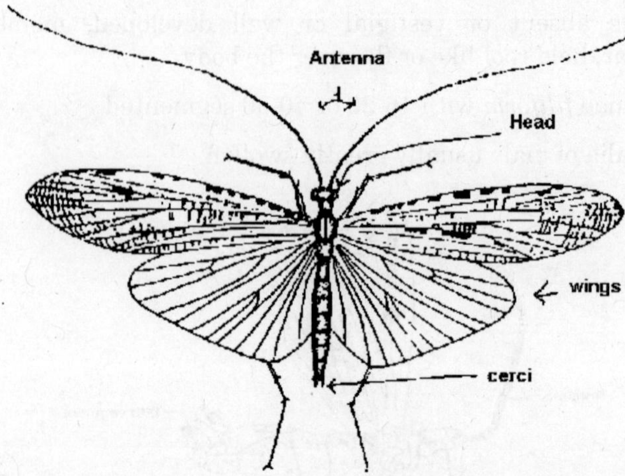

Fig. 3.34a. Trichoptera (Caddis Fly).

Habit and Habitat

1. Nocturnal, Larvae are aquatic.
2. Eggs are deposited in gelatinous mass attached to roots, stem or other submerged objects in fresh water streams. The larva construct cases by webbing together sand particles, small stones, sticks etc. They remain within the case and retract with in if danger threatens.
3. Larva feed on diatoms, algae or are carnivorous and predate on the larvae of chironomus, simulium, crustaceans etc.

Collection

By aerial netting. Water side branches can be disturbed with stick. By turning stones along streams in hilly regions. Light trap during night is a good source of collection.

ORDER-LEPIDOPTERA (Butterflies and Moths)

1. Body covered with overlapping flat scales.
2. Mouthparts modified into a coiled sucking proboscis.
3. Wings 2 pair, clothed with scales. Forewings are always larger than hind wings.

4. Antennae long and slender, clubbed, clavate, pectinate or hooked, plumose in males of moths.
5. Holometabolous insects.

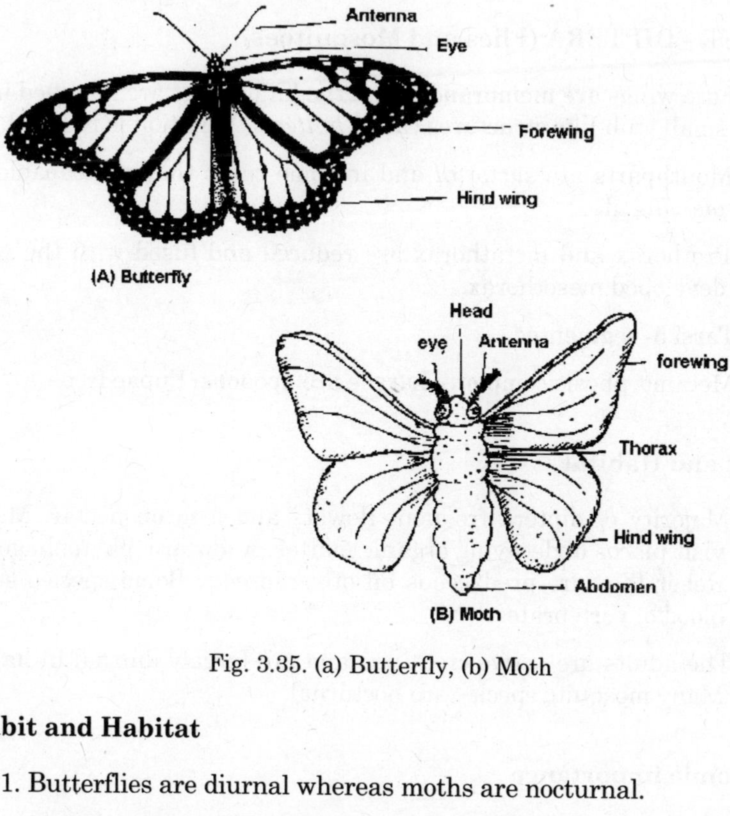

Fig. 3.35. (a) Butterfly, (b) Moth

Habit and Habitat

1. Butterflies are diurnal whereas moths are nocturnal.
2. Adult lepidoptera feed on nectar from flowers and juices from fruits. Larvae have *biting mouthparts* and are phytophagous feeds on leaves, stem and grains.

Economic importance

1. Lepidoptera are of great economic importance because of the damage caused by their larvae to the commodities of human interest.
2. Many of them play a useful role in pollination of flowers.
3. Silkworm moth yields silk which is of high commercial value.

Collection

Collect with the help of *insect net and pin in the box.* Larvae can be preserved in alcohol. Larvae can also be collected from the host plants.

ORDER - DIPTERA (Flies and Mosquitoes)

1. Fore wings are membranous and the hind wings are modified into small stub-like structures called *halteres.* Venation is reduced.
2. Mouthparts are *suctorial* and in some cases they are capable of *piercing* also.
3. Prothorax and metathorax are reduced and fused with the well developed mesothorax.
4. Tarsi 5- segmented.
5. Metamorphosis complete. Larvae are apodous. Pupae free.

Habit and Habitat

1. Majority of diptera frequent flowers and feed on nectar. Many visit places of decaying organic matter. A few are phytophagous. Rober flies are predaceous on other insects. Some species suck blood of vertebrates.
2. The adults are active creatures and are largely diurnal in habit. Many mosquito species are nocturnal.

Economic importance

1. They cause great annoyance.
2. Blood sucking insects are capable of transmitting disease causing micro-organisms from one host to other.

Tse-tse fly act a vector for *Trypanosoma* which causes sleeping sickness in Africa. Mosquito transmits yellow fever, dengue, malaria, elephantasis. House flies spreads cholera.

Hippobosca is parasitic on dogs and cattles.

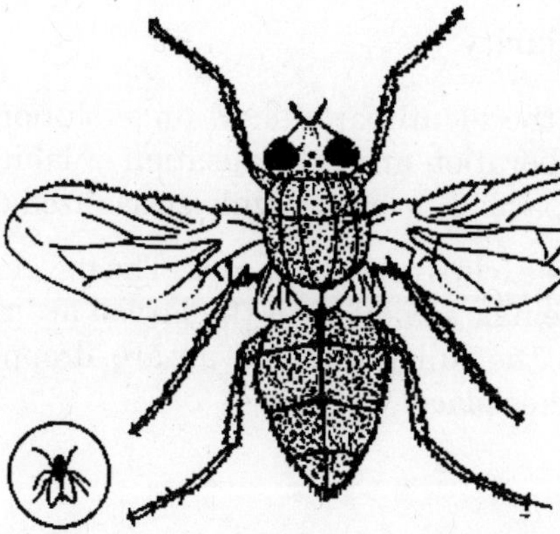

(a) Housefly (*Musca domestica*)

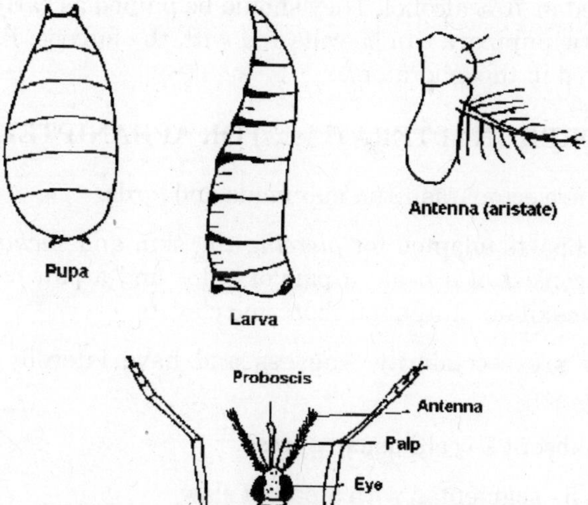

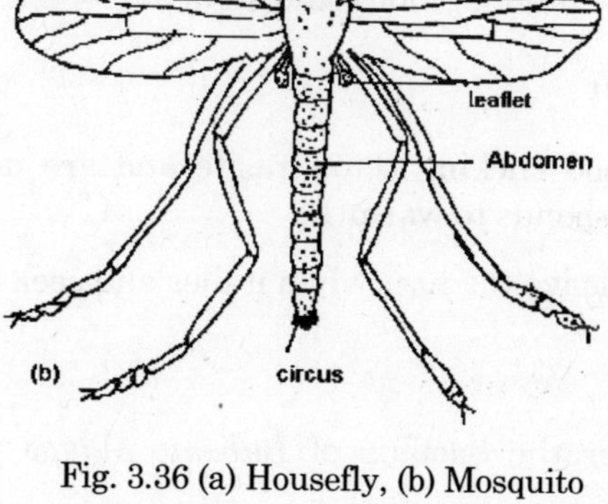

Fig. 3.36 (a) Housefly, (b) Mosquito

Biological peculiarity

1. In Diptera, the mouthparts show an evolutionary trend leading towards elaboration and specialisation of labium with the atrophy of other parts. It is most highly specialized of insect orders.

2. *Pupipara* are exclusively viviparous. The development takes place inside the female's body until the larvae are ready for pupation. As soon as the full grown larvae are dropped to the ground, pupation takes place.

Collection

Collect with the help of insect net and pin them. Prepare a permanent slide of mouthparts and other immature stages. Larvae can be preserved in 70% alcohol. They should be pinned as early as possible Ectoparasitic pupipara can be collected with the forceps. Parasitic flies can be reared in the laboratory.

ORDER - SIPHONAPTERA (Fleas) OR APHANIPTERA

1. They are ectoparasite on mammals and birds.

2. Mouthparts adapted for *piercing* the skin and *sucking* the blood and consist of a beak, a pair of palps and a pair of short blade like maxillae. Antennae short and clavate.

3. They are secondarily wingless and have laterally compressed body.

4. Eyes absent 2-ocelli usually present.

5. Tarsi 5 - segmented with a pair of claw.

6. Legs *saltatory*, hind legs long and stout.

Habit and Habitat

1. They are blood sucking ectoparasite and are negatively phototactic and responds to warmth.

2. They at once leave the host when it dies and seek fresh host.

Economic status

1. They transfer the bacillus of *Bubonic plague* from rodents to humans.

2. Occasionally they are pest of domestic fowl. *Tunga penetrans* causes painful swelling in the feet of man beneath the skin, out of which protrudes the abdomen of flea.

3. *Xenopsylla cheopis* infest house rats and transmits plague bacilli (*Yercinia pestis*) to man.

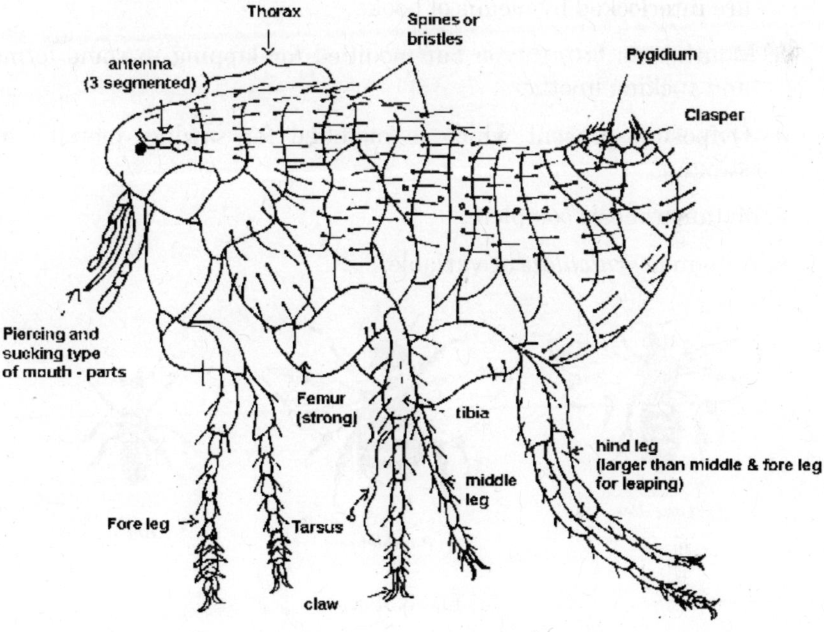

Fig.3.37. *Xenopsylla cheopis*

Biological status

Fleas are highly modified. Absence of mandibles and presence of *suctorial mouthparts* suggests a possible origin from Diptera or Mecoptera and their larvae strongly resembles to the immatures of dipterous sub order *Nematocera.*

Collection

Prepare a permanent slide. Examine the skin of a friendly street dog when the animal is a sleep. Collect the fleas with help of forceps or fingers and dropped into 70% alcohol. Freshly anaesthetized rat can also be used for the collection of fleas.

ORDER - HYMENOPTERA (Bees, Wasps and Ants)

1. Narrow waist (*petiole*) between thorax and abdomen present.
2. Two pair of membranous wings with reduced venation. Hind wings are smaller than the fore wings. The hind margin of the fore wings and anterior margin of hind wings of the same side are interlocked by means of hooks.
3. Mouthparts *biting* type but modified for lapping in some forms and sucking in others.
4. Ovipositor present which is modified for sawing, piercing or stinging.
5. Metamorphosis complete.
6. Antennae *geniculate* or variable.

Fig. 3.38a.

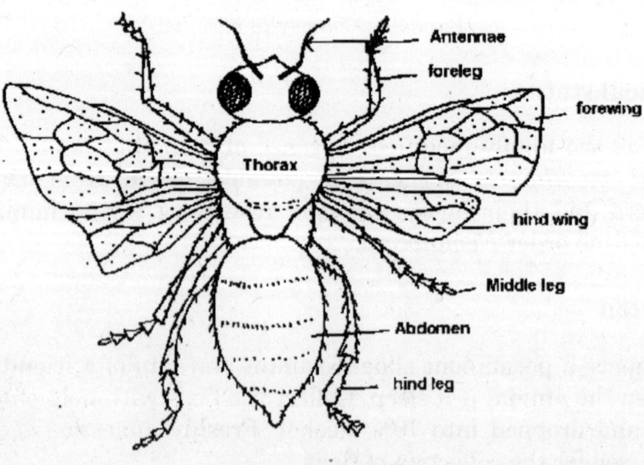

Fig. 3.38b. *Apis*

Habit and Habitat

1. They are *diurnal* insects who love sunshine and warmth. They lead a predatory life or gather pollen and suck the sweet juices of plants.
2. They are free living or phytophagous, entomophagous parasitic or social insects.

Economic importance

1. They are the agents of *biological control* of many crop pests. They parasitises many lepidopterous pests.
2. Honeybee is very beneficial as it provide honey and bees wax.
3. Many are crop pests e.g. Saw flies.
4. They are helpful in pollination of various plants. Fig wasps (chalcidae) pollinates *Ficus* genera.

Biological peculiarity

1. Bees and ants exhibits highly organised social life and polymorphism.
2. They exhibit *parasitism*. Braconids, Chalcids, Eupelmids, Bethylids etc. are parasitic on caterpillars.

Honey bee (Apis)

It lives in colony and is a social insect.

The colony consists of *queen, workers* and *drones*. The workers are the largest in numbers in the colony and do all the duties relating to food collection, bringing nectar, secreting wax, looking after the immatures, building and cleaning the hives etc. The queen lays the eggs whereas the drones only copulate with the queen.

The larvae to whom the royal jelly is given (process called *trophallaxis*) develops into queen. Rock bee, *Apis dorsata*, little Indian bee, *Apis florea*, European bee, *Apis mellifera* and Indian bee, *Apis indica* are commonly found in India. Bee keeping is called *Apiculture*. They are of great economic importance as they are the source of honey and bees wax. They also pollinate large number of flowers. Honey bees are the insects known for *air conditioning* of the hive. During summers they wet their wings and fans the hive as a result the temperature of the hive reduces and in winters they crawl and move around within the hive, this increased activity enhances the temperature of the hive.

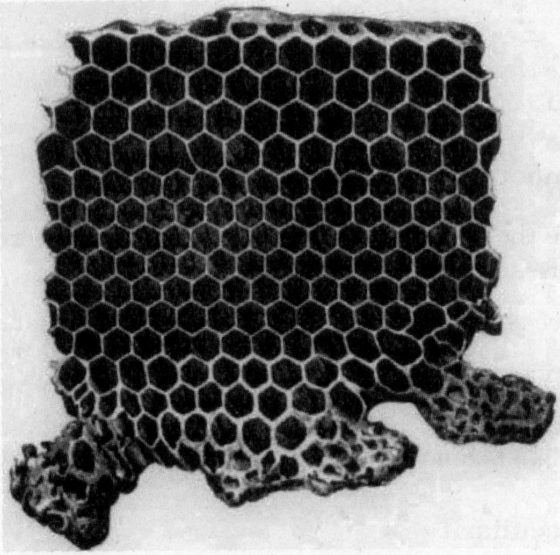

Photo: A part of Honey comb

Wasps (Vespidae)

The wasps feeds on caterpillars, mantids, bugs, grasshoppers, beetles (some on fruit juices), sweets, sugar. They generally made the nests of papery material consisting of chewed vegetable fibre. The fore tibia bears a cleaning comb through which the antennae is drawn. Middle tibia is provided with two terminal spurs and wings are longitudinally folded in repose. They are of great economic importance as they feeds on the larvae of many insect pests. The *hornets* and *vespa* if disturbed are painful due to their stings.

Collection

They can be collected with the help of insect net. Pin in the box. They can also be preserved in 90% alcohol. Nest of different wasps can also be collected. Prepare permanent slide of mouthparts, antennae, wings and pollen basket. They can be collected by hand picking, sweeping, light trap and Berlese funnel. They are best preserved dry. Smaller specimens are mounted on small triangular cards.

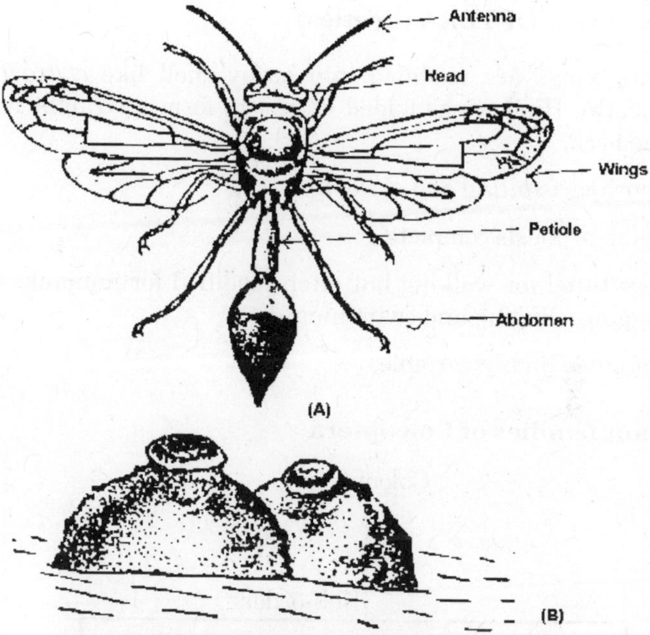

Fig. 3.39. (A) *Eumenes* sp. (B) Nest

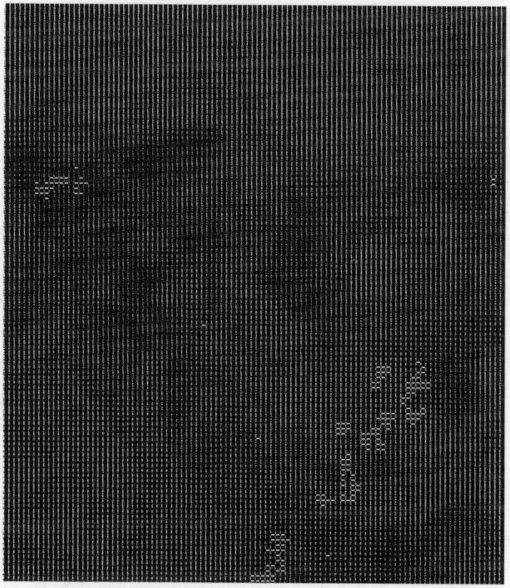

Photo: Sphecid wasp on its nest

ORDER - COLEOPTERA (Beetles)

1. Fore wings are modified into horny shell like *elytra* (veinless sheath). Hind wings folded at rest to form a straight line down the back.
2. Mouthparts *biting* and *chewing*.
3. Metamorphosis complete.
4. Legs fitted for walking but often modified for jumping, running, digging, clinging and swimming.
5. Antennae highly variable.

Important families of Coleoptera

Coleoptera

Sub-order

Adephaga

Family

1. Cicindelidae
2. Carabidae
3. Dytiscidae
4. Gyrinidae

Polyphaga

Family

1. Scarabaeidae
2. Hydrophilidae
3. Staphylinidae
4. Buprestidae
5. Elateridae
6. Lampyridae
7. Dermestidae
8. Coccinellidae
9. Tenebrionidae
10. Meloidae
11. Cerambycidae
12. Chrysomelidae
13. Curculionidae
14. Anobiidae
15. Bruchidae

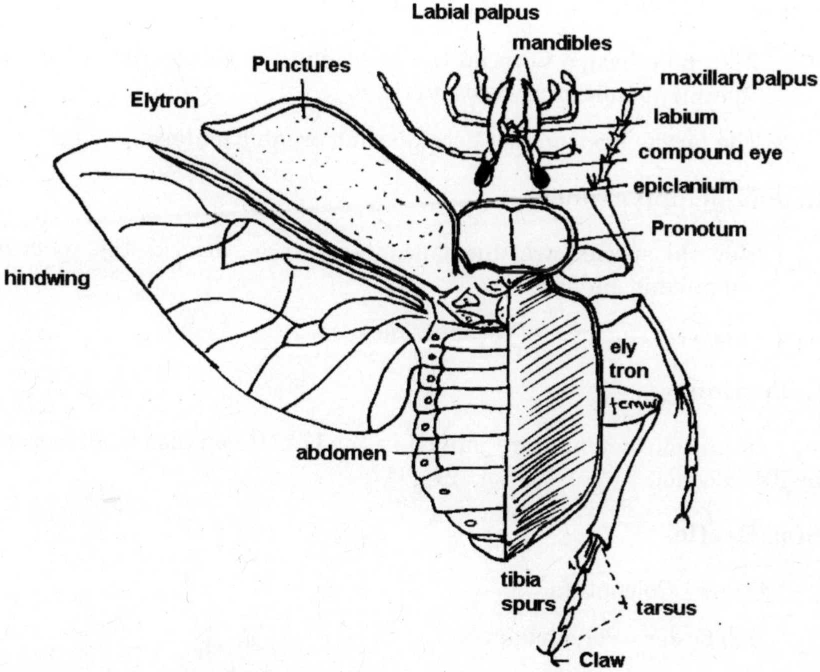

Fig. 3.40. Coleoptera : Generalized plan.

Rhinoceros Beetle (*Oryctes*)

Coleoptera

Polyphaga

(i) first-abdominal sternite not divided by the hind coxal cavities.

Scarabaeoidea

(i) Antennae 10 segmented.

Scarabaeidae

(i) Forelegs with spines or toothed.

(ii) Hind tibia with a single terminal spur.

(iii) Abdomen not wholly completely covered with elytra.

Oryctes rhinoceros

1. The male bear a horn on the head like the rhinoceros hence the common name *rhinoceros beetle.*
2. The larvae lives amidst decomposing organic material.

Economic importance

1. Several species are injurious, their larvae attacks the roots of sugarcane and rice.
2. It is a serious pest of coconut trees.

Collection

Hand collected and are pinned in the box. It can also be preserved in 70% alcohol.

Stag Beetle

Order - Coleoptera

Sub Order - Polyphaga

Super family - Scarabaeoidea

Family - Lucanidae

(i) Antennae elbowed, rarely straight (ii) Mandibles strongly developed in males and prominent.

Genus - Lucanus

1. Commonly called *Stag beetle.*

Habit and Habitat

1. Adults are often attracted to light at night.
2. They are found in woods, sandy beaches.
3. They feed on the juices of decaying vegetation.

Collection

Hand collected and are pinned in the box. They can also be preserved in 70% alcohol.

Jewel Beetle

Family - *Buprestidae*

1. They are commonly called metallic wood borers.
2. Antennae serrate 11 segmented.
3. Thorax and abdomen firmly united.
4. Adults love bright sunshine, inhabit wooded areas.
5. Larvae bore into the stems.
6. Eyes absent.
7. Elytra strongly sclerotised and hard.

Habit and Habitat

They prefer wooded area. In Rajasthan, they have been seen feeding on the leaves of '*Prosopis cinereria* and Ber plants.

Economic Importance

1. Buprestid Beetles have been utilized in embroidery and other works of art in India and other countries.
2. They are sold as curiosities by street hawkers in hills for their irridiscent colours of elytra.
3. They are also used as inlay in gem and jewellery.
4. Brilliant green species from India (*Sternocera*) China and Japan is much sought by jewellers. *Chrysochroa mutabilis* is very fascinating.

Biology

Life cycle is completed in one year. Adults are seen from July to October in Rajasthan. The eggs are laid after the mating occurs. The eggs are hard and nearly equals to the size of pea. They undergo dormancy upto 7-8 months.

Collection

They can be easily collected by picking with hands or nets.

Fig. 3.40. Jewel beetle

(a) Stage beetle (b) *Rhinoceros beetle*

Fig. 3.40 (a,b).

Aquatic Beetles

Hydrophilus

Family : Hydrophilidae

Note: *Hydrophilus* bears superficial resemblance with *Cybister,* the difference lies in antenna.

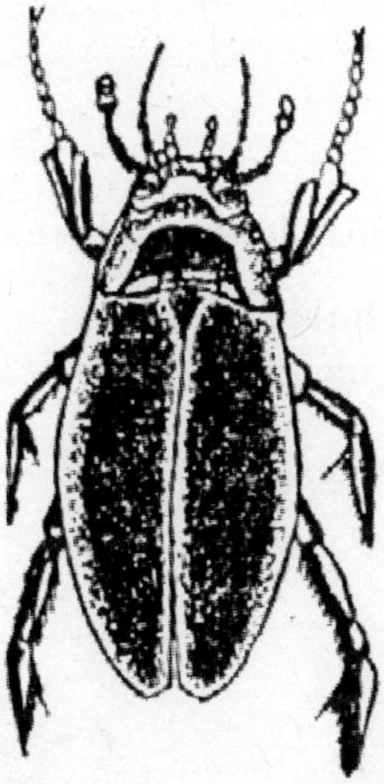

Fig. 3.40(c) *Hydrophylid* beetle

Habit and Habitat

Found in fresh water ponds and lakes. Adult is vegetarian and larvae are predaceous. Adult is active swimmer.

Characters

1. Beetle is shiny black in colour and measures about 30 mm. in length.
2. Antennae ends in a club. It lies below the head.
3. Maxillary palps are longer than antennae.
4. Hind legs adapted for swimming (*natatorial*).

Examples. *Hydrophilus olivacous, Hydrophilus indicus Berosus* is a small species found in tanks and pond.

Peculiarity: (1) The mode of respiration under water. (2) The female secrete cocoon for egg laying.

They are hand collected. They often come to light in rainy season. They are preserved dry in the inset boxes or also into 70% alcohol.

Cybister tripunctatus

Family - Dytiscidae

Habit and Habitat

1. Large size aquatic beetle found in lakes and pond.
2. It is predaceous in habit. Larva is aquatic and carnivorous. It feeds voraciously on mollusca, tadpoles and insects.

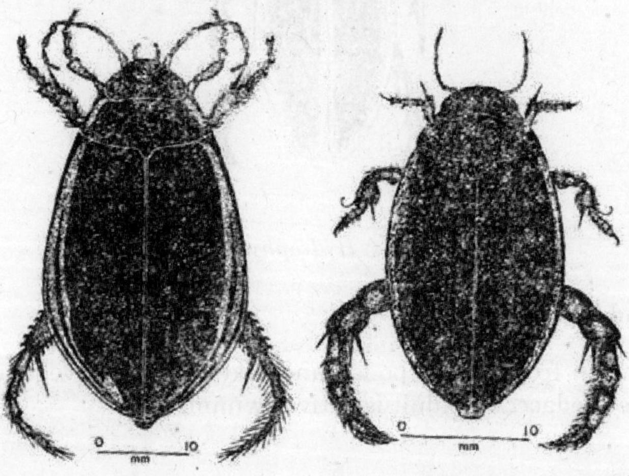

Fig. 3.40(d) *Cybister limbatus* Fabr.; *Cybister rugulosus* Redt.

Characters

1. Antennae filiform.
2. Measures about 45 mm in length and is black in colour with a lateral brown strip.
3. Hind legs adapted for swimming (*natatorial*).
4. In males the basal three tarsal segments of the fore legs are dilated.

Collection

Hand collected often come to the light in rainy season. Preserved dry and pin them in box.

Economic importance

The larva of *Eretes stictus* is said to feed on mosquito larva thus helpful in Biological control of mosquito.

Biological peculiarity

The adult uses air stored under elytra for respiration. It comes periodically to the surface of the water for fresh supply of air.

The larva comes to the surface of the water and breathes atmospheric air by means of a pair of caudal spiracles.

Lady Bird Beetle (*Chilomenes*)

Family - Coccinellidae

Habit and Habitat

1. It feeds voraciously on different species of aphids and found in association with aphid colonies.

Characters

1. It is oval yellowish red in colour. There are 6 Black zig-zag (wavy) stripes and spots on the body (see figure on page 101).
2. The larva is dark and active.

Economic Importance

Both adult and larva consumes large amount of aphids thus are potent agent of Biological control of aphids. It controls the natural population of aphids.

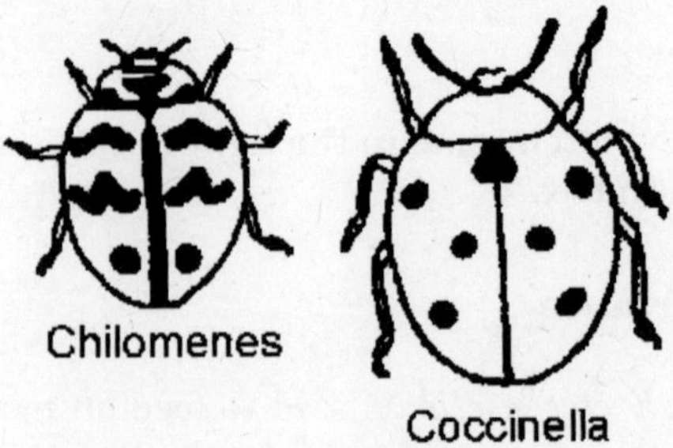

Figure 3.40(e). Lady Bird Beetles

Collection

Hand collected from the crop fields, garden plants etc. the adults are pinned in boxes and larvae can be preserved in 70% alcohol in glass vials. 7 spotted lady bird beetle is *Coccinella septempunctata.*

Blister Beetle (*Mylabris pustulata*)

Family - Meloidae

(The members of family - are called Blister beetles for the reason they yield pharmaceutical product *Cantharidin).*

Habit and Habitat

It generally feeds on flowers and tender shoots of garden plants, pulses and various other plants, however, it is not regarded as serious crop pest.

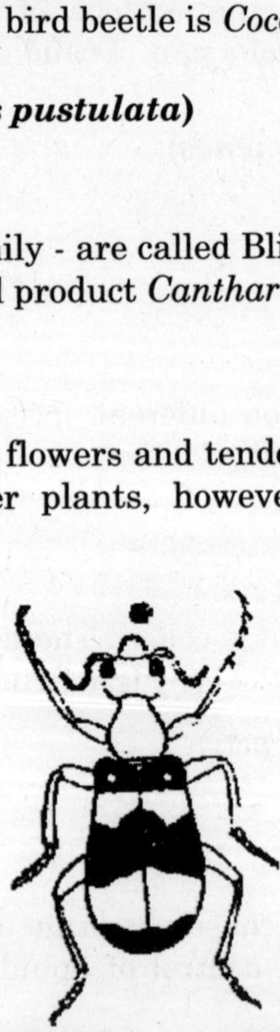

Figure 3.40(f). Blister Beetle.

Characters

1. It is bright red or orange with black bands.
2. Head deflexed, neck narrow and claws are serrate.

Biological peculiarity

1. They exhibit *hypermetamorphosis* which is distinguished by the occurrence of two or more distinct larval types. The newly hatched larva from the egg, is known as *triangulin.* They feed on the eggs of orthoptera or hymenoptera like bees. This stage followed a soft-bodied eruciform larva. This passes into pseudo-pupal stage. Again there is a further larval instar which is succeeded by pupa.
2. It exude oily yellow orange liquid droplet (from the joints of the legs) which causes painful blisters on the human skin.

Economic Importance

1. All blister beetles yields a pharmaceutical product, oily in nature called *cantharidin.* It is regarded as hair restorer. The powder of dried blister beetle is used to extract cantharidin for the manufacture of hair oil and medicines.

Biology

It lay eggs in the soil before the cold weather and the larvae develop very slowly. It devours the eggs of orthoptera.

Collection

Hand collected, dried and pinned in the box.

ORDER- STREPSIPTERA (Stylops)

1. Mouthparts *atrophied.*
2. Fore wings modified as *halteres*, female apterous.
3. Legs without trochanter, claws often absent.
4. Antennae with 4-7 segments, some segments flabellate.

5. Dimorphic, free or endoparasitic.

Habit and Habitat

1. Short lived (male live for 1-2 days)
2. The adult never feeds.
3. Their larvae lives free within the body cavity of host and does not attack any organs.
4. They attack Orthoptera, Heteroptera, Homoptera and Hymenoptera.

Biological peculiarity

The members are internal parasites of other insects and cause *Parasitic castration* of the host.

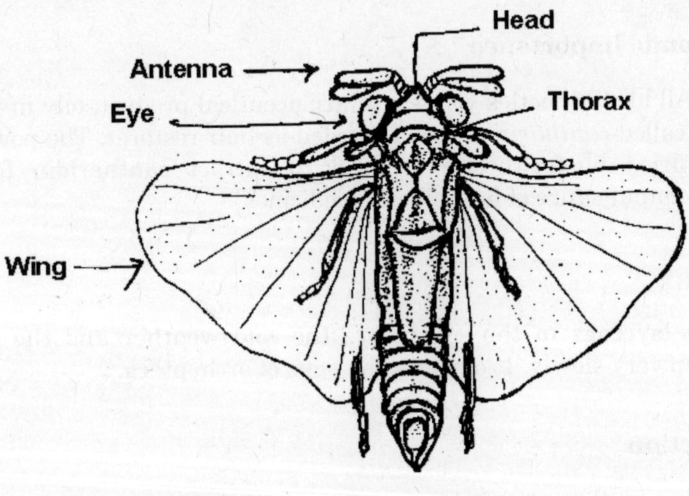

Fig. 3.41. Strepsiptera (*Stylops*)

Collection

The best method to collect them is to look in their hosts viz. Hymenoptera or other insects. Stylopised insects may be kept in captivity for obtaining males. Males sometimes found at light. They can be preserved in 90% alcohol.

CHAPTER - 4

IDENTIFICATION OF INSECTS WITH THE HELP OF TAXONOMIC KEYS

TAXONOMIC KEYS

The main purpose of the key is to facilitate identification and are most commonly used methods of identification. It is a printed information retrieval system into which are puts information regarding a specimen in hand and from which one gets an identification of the specimen to whatever level the key is designed to reach, for example upto order, family or species. A good key is dichotomous, not having more than two alternatives at any point. The language of the key is telegraphic. The phrase in each alternative are usually separated by semi colons.

Though the keys are necessary tools for identification yet they exhibit some limitations e.g. a key made for adults shall not work for young ones (larvae or nymphs). In adults if it for males, it will not work for females etc. The keys for young ones e.g. aquatic nymphs of odonata are separate. The younger stages are identified by following the relevant keys.

KEY TO ORDERS

1. Wings present....... 2

– Wings vestigial or absent.... 32

2. Fore wings horny, leathery or parchment-like at least at base, differing in texture from the hind wing, which is membranous.... 3

– Fore wings membranous.... 12

3. Fore wings modified as tegmina or hemelytra, containing veins..... 4

– Fore wings modified as veinless, horny elytra, hind wing when present folded crosswise and hidden under the elytra... 11

4. Mouthparts forming a segmented beak, adapted for piercing and sucking liquid... 5

– Mouthparts with mandibles adapted for biting and chewing solids.... 6

5. Head usually horizontal, with the beak arising from below the projecting downwards, the gula well developed, fore wings usually overlapping one another and lying flat over the abdomen when at rest..... .....***Heteroptera***

– Head usually vertically, with the beak arising from the back part so as to project backward between the fore legs, gula absent or represented by a membrane, fore wings not different from hind wings ....***Homoptera***

6. Hind wings not folded, similar to fore wings, social insects, living in colonies..... ....***Isoptera***

– Hind wings folding fan-like, broader than fore wings... 7

7. Antennae long and thread-like, prothorax large and free from mesothorax, cerci present, fore wings rarely minute, usually long. 8

– Antennae short, with few segments, at least one segment bearing a long thread-like arista, cerci absent, fore wings minute, minute species, parasitic in wasps and bees..... ....***Males of Strepsiptera***

8. Hind femora not larger than fore femora, body more or less flattened, with the wings superposed when at rest, tergites and sternites usually equal.. 9

– Hind femora almost always much larger than fore femora, and adapted for jumping, if not the fore legs adapted for burrowing, body more or less cylindrical, the wings held slopping against the sides of the body when at rest, tergites usually larger than sternites...... ....***Orthoptera***

9. Body elongate, head not concealed from above by the prothorax..... 10

– Body oval, much flattened, head usually concealed from above by the prothorax...... ....***Blattaria***

10. Prothorax much longer than mesothorax, fore legs usually heavily spined beneath and adapted for capturing and holding prey, cerci usually with several segments.... ....***Mantodea***

– Prothorax short, fore legs not raptorial, cerci uniarticulate ....***Phasmida***

11. Abdomen apically with movable, heavily sclerotized forceps, antennae long and slender, fore wings short, leaving most of the abdomen uncovered, hind wings nearly circular, delicate, radially folded from near the centre..... .... ***Dermaptera***

– Abdomen not ending in forceps, antennae variable, fore wings usually completely covering the abdomen...... ....***Coleoptera***

12. Two pairs of wings present.. 13

– Only one pair of wings present. 30

13. Wings long, narrow, fringed with long hairs, almost veinless, tarsi uni-or biarticulate, with swollen tips mouthparts asymmetrical, without biting mandibles, adapted for lacerating and sucking, plant tissues, no cerci, minute species ....***Thysanoptera***

– Wings broader and often with veins, if rarely linear, the tarsi have more than two segments and the last tarsal segment is not swollen. 14

14. Hind wing with the anal area folded in fan-like plaits when at repose, larger than the fore wing. 15

– Hind wings not folded, not larger than the fore wings, the anal area small and not separated..... 17

15. Tarsi with 5 segments, cerci inconspicuous... 16

– Tarsi with 3 segment, cerci segmented... ....***Plecoptera***

16. Costal area with few cross-veins, wing surface hairy, prothorax small. ....***Trichoptera***

– Costal area with many cross-veins, prothorax large............. ***Megaloptera***

17. Antennae short, inconspicuous, wings net-veined with numerous cross-veins, larvae aquatic.... 18

– Antennae larger, distinct, if rarely small, cross-veins few, larvae almost always terrestrial.. 19

18. Hind wing much smaller than fore wing, abdomen ending in long thread-like caudal filament and cerci, tarsi usually with four or five segment; short-lived, sluggish fliers.. ....***Ephemeroptera***

– Hind wing nearly like the fore wing, no caudal filament, tarsi triarticulate, long-lived, vigorous active.. ....***Odonata***

19. Head produced into a mandibulate beak, male genitalia usually greatly swollen and forming a reflexed bulb.... ....***Mecoptera***

– Head not produced into a mandibulate beak, male genitalia not as described above... 20

20. Mouth mandibulate...... 21
– Mouth haustellate, the mandibles not adapted for chewing solid food .. 28

21. Tarsi with 5 segments, if rarely with 3 or 4, hind wings smaller than fore wings..... 22
– Tarsi with 2, 3 or 4 segments, veins and cross-veins not numerous..... ... 25

22. Prothorax small or moderately long; if long, the fore legs are raptorial..... 23
– Prothorax long, cylindrical, much longer than head; fore legs normal... ....***Raphidiodea***

23. Wings similar, with many veins and cross-veins, prothorax more or less fress; if with reduced venation, wings covered with a powdery bloom.... 24
– Hind wings smaller than fore wings; costal cells without cross veins; prothorax fused with mesothorax.... ....***Hymenoptera***

24. Costal cell at least in fore wing with many cross-veins.. ***Neuroptera***
– Costal cells without cross-veins.... ....***Mecoptera***

25. Wings equal, rarely the hind wing larger, held superposed on top of abdomen when at rest; M fused with Rs for a short dista-nce near the middle of wing; tarsi with 3, 4 or 5 segments.... 26
– Hind wings smaller than fore wings; wings at rest folded back against the abdomen; R and M not fused; tarsi with 2 or 3 segments.... 27

26. Tarsi with 3 segments, fore metatarsus swollen; cerci conspicuous; usually solitary species.. ....***.Embioptera***
– Tarsi apparently with 4 segments, fore metatarsus not swollen; cerci usually minute; social insects... ....***Isoptera***

27. Cerci absent; tarsi with 2 segments; Rs and M branched .... ***Psocoptera***
– Cerci present; tarsi with 2 segments; wings shed at maturity; venation greatly reduced, Rs and M simple... ....***.Zoraptera***

28. Wings not covered with scales, not outspread when at rest; prothorax large; antennae with few segments; mouthparts forming a segmented beak. 29
– Wings and body covered with scales forming definite coloured patterns; antennae many segmented; mouthparts when present forming a long coiled proboscis.... ....***.Lepidoptera***

29. Beak arising from the back of head.... ....***Homoptera***

– Beak arising from the front of head.. ....***Heteroptera***

30. Mouth non-functional; abdomen with a pair of caudal filaments..... 31

– Mouthparts forming a proboscis, only rarely vestigial; abdomen without caudal filaments.... ....***Diptera***

31. No halteres; antennae inconspicuous; cross-veins abundant.... ....***Ephemeroptera***

– Hind wings reduced to hook like halteres; venation reduced to a forked vein; cross-veins absent; minute delicate males of scale-insects... ....***Homoptera***

32. Body with more or less distinct head, thorax and abdomen and segmented legs and capable of locomotion... 38

– Without distinct body parts, without segmented legs; incapable of locomotion.... ....***Homoptera***

33. Free-living insects.. 34

– Parasitic on warm-blooded animal.... 51

34. Mouthparts vestigial, retracted within the head; underside of the abdomen with styles or other appendages... 35

– Mouthparts not vestigial and not restricted within the head.. 38

35. Abdomen with 10 or 12 segments, without ventral suckers and springing apparatus. 36

– Abdomen with 6 or less segments, with a forked sucker at base below and usually with a springing apparatus near tip... ......***Collembola***

36. Basal three segments of abdomen with styles; antennae absent; head pear-shaped; prothorax short.... ....***Protura***

– Ventral styles occurring up to the seventh segments on the abdomen; antennae thread-like; prothorax not short.. 37

37. Body never scaly; mouthparts concealed within the head except for the palpi; tip of abdomen without a median process ....***Aptera (Diplura)***

– Body usually covered by scales; mouthparts not sunk within the head; abdomen apically with a cerciform appendage.... ....***Thysanura***

38. Abdomen apically with strong, movable forceps; prothorax free. ....***Dermaptera***

– Abdomen not ending in forceps.. 39

39. Abdomen strongly constricted at base; prothorax fused with mesothorax.. ...*Hymenoptera*

– Abdomen not strongly constricted at base; broadly attached to the thorax.. ...40

40. Head produced into a mandibulate beak. ...*Mecoptera*

– Head not produced into a mandibulate beak... ...41

41. Prothorax inconspicuous; very small jumping species louse- like in appearance.... ...*Psocoptera*

– Prothorax larger; larger species... ...42

42. Hind femora enlarged; hind legs adapted for jumping; wing- pads of larvae, when present, in inverted position, the metathoracic overlapping the mesothoracic.... ...*Orthoptera*

– Hindlegs not adapted for jumping; wing-pads, when present, in normal position.. ...43

43. Prothorax much longer than mesothorax; fore legs raptorial.... ...*Mantodea*

– Prothorax normal... ...44

44. Cerci present; antennae usually with more than 15 segments ...45

– Cerci absent; antennae usually with 11 segments ...*Coleoptera*

45. Cerci with more than three segments.. ...46

– Cerci with 1 to 3 segments.. ...48

46. Body flattened and oval; head concealed under the oval prothorax... ...*Blattaria*

– Body elongate; head horizontal.. ...47

47. Cerci long; ovipositor sclerotized, exserted; tarsi with 5 segments.. ...*Grylloblattoidea*

– Cerci short; no ovipositor; tarsi with 4 segments... ...*Isoptera*

48. Tarsi with 5 segments; body slender and stick- like.... ...*Phasmida*

– Tarsi with 2 to 4 segments; body not linear... ...49

49. Fore metatarsus swollen.... ...*Embioptera*

– Fore metatarsus normal.... ...50

50. Tarsi apparently with 4 segments; cerci multiarticulate; antennae with 9 to 30 segments... ...*Isoptera*

– Tarsi with 2 segments; cerci uniarticulate; antennae with 9 segments.... .... ***.Zoraptera***

51. Body depressed.... .... .52

– Body compressed...... .... ***Aphaniptera***

52. Mouthparts mandibulate.... 53

– Mouthparts haustellate.... 55

53. Cerci long; mouth inferior.... 54

– Cerci short; mouth anterior, generally elongate insects with somewhat triangular head; parasites of birds or mammals... .... ***Pthiraptera***

54. Cerci strongly bent or angulate in the middle; eyes present; antennae nearly as long as body; ectoparasites of bats... .... ***.Dermaptera***

– Cerci straight; eyes absent; antennae short; ectoparasites of rodents..... .... ***Diplogossata***

55. Antennae inserted in pits, not visible from above pupipara... .... ***Diptera***

– Antennae not as above..... .... .56

56. Beak unsegmented; tarsi modified as hooks for grasping hairs of host; permanent ectoparasites of mammals.... .... ***Pthiraptera***

– Beak segmented; tarsi not hooked; temporary parasites..... .... ***Heteroptera***

KEY TO HYMENOPTERA (Important families only)

1. Slender waist or petiole absent.... .... *S.O.* ***Symphyta,*** 2

– Petiole present. ....*S.O.* ***Apocrita,*** 4

2. Antennae clubbed.. ....(Cimbicid sawflies) ***Cimbicidae***

– Antennae filiform.. 3

3. Usually 1 inch or more in length... ....(horntails) ***Siricidae***

– Usually less than 1/2 inch in length.... ....(sawflies) ***Tenthredinidae***

4. Wing venation reduced to a single anterior vein.... ....(Chalcid wasp) ***Chalcidae***

– Not as above...... 5

5. Antennae long with more than 16 segments... 6

– Antennae with fewer than 16 segments... 7

6. One recurrent vein.... ....(braconids) ***Braconidae***

– Two recurrent veins.... ....(ichneumonids) ***Ichneumonidae***

7. Pronotum extending posterior to tegula, may be wingless, hairy and slender.... 8

– Pronotum not extending, winged, stout bodied... 11

8. Segment in petiole with a dorsal bump or spine.. ....(ants) ***Formicidae***

– Petiole without bumps or spines.... 9

9. Usually covered by numerous hair.... .... (velvet ants) ***Mutillidae***

– Usually little conspicuous hair.. 10

10. Mesopleuron divided by suture, hind femora longer than abdomen,. ....(spider wasps) ***Pompilidae***

– Not as above, wings folded longitudinal when flexed. .... (vespids) ***Vespidae***

11. With branched and plumose hair present ....(bees), 12

– Without the above ....(sphecid wasps) ***Sphecidae***

12. Thorax usually metallic green... ....(mining or sweet bees) ***Halictidae***

– Thorax not metallic green. 13

13. Dense hair (yellow) only on 1st adominal segment, rest of abdomen black or metallic blue. ....(carpenter bees) ***Xylocopidae***

– Dense hair on more than 1st abdominal segment ....(apid bees) ***Apidae***

KEY TO HEMIPTERA (Sub-order Heteroptera) (Important families only) (For details see Appendix III)

1. Hindleg without tarsal claws.. 2

– Hindleg with tarsal claws 3

2. Forelegs spoon-shaped ....(water boatmen) ***Corixidae***

– Forelegs not spoon-shaped, clawed............ (back swimmer) ***Notonectidae***

3. Antennae easily seen, not reduced.. 6

– Antennae reduced and difficult to locate.... 4

4. Abdomen with long slender caudal appendage .... (water scorpion) ***Nepidae***

– Abdomen lacking long slender caudal appendage... 5

5. Body as wide as long, rough exoskeleton, 2 ocelli.... .... (toad bugs) ***Gelastocoridae***

– Body much longer than wide, no ocelli ....(giant water bugs) ***Belostomatidae***

6. Longs narrow legs, hind femur extend much beyond abdomen tip ....(water spiders) ***Gerridae***

7. Wingless or forewings reduced to short scales ....(bed bugs) ***Cimicidae***

– Wings fully developed.. 8

8. Pronotum with broad flanges, lace-like pattern in wings and pronotum ....(lace bugs) ***Tingidae***

– Not as above... 9

9. Loop veins in membrane of hemielytra... 10

– No loop veins in membrane of hemielytra.. 12

10. Femur of forelegs very thick, triangular-shaped... ....(*ambush bugs*) ***Phymatidae***

– Femur of forelegs not thick or triangular.... 11

11. Beak (proboscis) curved, 3-segmented, forewings not tilted downwards.. ....(assasin bugs) ***Reduviidae***

– Beak essentially straight, 4-segmented, forewings distinctly tilted posterior to abdomen.... ....(plant bugs) ***Miridae***

12. Scutellum greatly enlarged, antennae 5-segmented.... ....(stink bugs) ***Pentatomidae***

– Scutellum not enlarged... 13

13. Forewings membrane with 4-6 veins... ....(Lygaeid bugs) ***Lygaeidae***

– Forewings membrane with more than 10 veins... ....(coreid bugs) ***Coreidae***

KEY TO HEMIPTERA (Sub-order Homoptera)
(Important families only)

1. Setaceous antennae..... 2

– Filiform antennae..... 6

2. Over 1/2 inch, 3 ocelli...... .... (cicadas) ***Cicadidae***
– Less than 1/2 inch, 2 ocelli or more.... 3
3. Pronotum enlarged dorsally... ....(tree-hoppers) ***Memberactidae***
– Pronotum not enlarged dorsally.. 4
4. Antennae arise beneath or behind
eyes ....(plant-hoppers) ***Fulgoridae***
– Antennae arise between eyes.. 5
5. Hind tibia with one or more rows of numerous
spines.. ....(leaf-hoppers) ***Cicadellidae***
– Hind tibia without such rows of spines.. ....(spittlebugs) ***Cercopidae***
6. Abdomen usually with cornicles, body
rounded ....(aphids) ***Aphididae***
– Abdomen lacking cornicles, body not rounded.... 7
7. Body covered by waxy scales.. ....(scales) ***Coccidae***
– Body not covered by waxy scales.... ....(jumping plant lice) ***Psyllidae***

KEY TO THYSANOPTERA (Sub-order Terebrantia)

Ovipositor saw like, abdomen tip conical in female and rounded in male, wing venation present.

1. Antennae 9-segmented, body flattened, ovipositor bend
upwards.... ....***Aeolothripidae***
– Antennae 6-10 segmented, body less flattened, ovipositor bent
downwards.... 2
2. Prothorax without sutures, wing-surface hairy... 3
– Prothorax with sutures, wing smooth... ....***Merothripidae***
3. Antennae 9-10 segmented, without apical stylets.... 4
– Antennae 6-8 segmented with apical stylets).... ....***Thripidae***
4. 3rd antennal segments cylindrical, not conical... .... ***Hemithripidae***
– 3rd and 4th antennal segments enlarged and
conical.. ....***Heterothripidae***

KEY TO THYSANOPTERA (Sub-order Tubulifera)

Ovipositor absent, abdomen tip tubular, venation absent in forewings.

1. Maxillary palps 2- segmented, antennae 8-segmented... 2
– Maxillary palps 1-segmented, antennae 4-7 segmented .... ***Urothripidae***
2. Last abdominal segment not tubular but swollen............. ***Pygothripidae***
– Last abdominal segment cylindrical or tubular... 3
3. 8th abdominal segment without peg-like projections... 4
– 8th abdominal segment with long peg-like projections.. ....***Chirothripoididae***
4. Tube much shorter than remaining abdominal segments taken together.... 5
– Tube about as long as remaining abdominal segments taken together.. ....***..Hystricothripididas***
5. 3rd antennae segment with prominent sensory rods..... .... ***Ecacanthothripididae***
– 3rd antennae segment with sensory rods as strong as on other segments..... 6
6. Antennae sense-cones extraordinarily long, eyes large and touching mouthcone pointed.... ....***Eupathithripidae***
– Internal sense-cones not unusually developed, eyes not touching.... ....***Phloethripidae***

KEY TO MALLOPHAGA (Sub-order : AMBLYCERA)

Antennae capitate, 4-segmented, concealed, meso-and metathorax separate:

1. Infesting birds. ....(*Menopon, Trinoton*) **Menoponidae**
– Infesting mammals... 2
2. Tarsi with 1 claw. ....(*Gyropus*) **Gyropidae**
– Tarsi with 2 claws..... ....(*Boopia, Heterodoxus*) **Boopiidae**

Sub-order : ISHNOCERA

Antennae filiform, 3-5 segmented, meso-and metathorax fused

1. Antennae 5-segmented, tarsi with paired claws, infesting birds... .... (*Lipeurus*) **Philopteridae**

– Antennae 3-segmented, tarsi single clawed, infesting mammals....... ....*(Trichodectes)*, **Trichodectidae**

Sub-order : RHYNCHOPHTHIRINA

Head prolonged into a rostrum, has only one family with one species .... *(Haematomyzus elephantis* found on Elephants) **Haematomyzidae**

KEY TO SIPHUNCULATA (ANOPLURA)

1. Body densely clothed with setae (scaly sometimes), abdomen without sclerotised plates, confined to carnivores. ....***Echinophithiriidae***

– Body less densely clothed, never on carnivores.. 2

2. Abdomen membranous except for genital parts.. 3

– Abdomen plates frequently present... 4

3. Only 1 pair of abdominal spiracles (in 8th segment) present.... ....***Neolinognathidae***

– 6 pairs of abdominal spiracles present.... .... ***Linognathidae***

4. Abdominal tergal and sternal plates distinct, paratergal plates projecting.... ....***Hoplopleuridae***

– Abdominal tergal and sternal plates absent, paratergal plates non-projecting.... 5

5. Abdominal cuticle unwrinkled, and membranous except in genital region..... ....***Pediculidae***

– Abdominal cuticle final wrinkles, plates sometimes indistinctly present.... ....***Haematopinidae***

KEY TO COLEOPTERA (Important families only)

1. Each eye divided into dorsal and ventral halves, 2nd and 3rd pair of legs short.. ....(Whirlinig beetles) ***Gyrinidae***

– Eyes not divided, legs short... 2

2. Legs modified for swimming.. 3

– Legs not modified for swimming.. 4

3. Sternal keel present ....(water scavengers) ***Hydrophilidae***

– Sternal keel absent. ....(predaceous diving beetles) ***Dytiscidae***

4. Head usually prolonged into a snout, antennae elbowed... ....(weevils) ***Curculionidae***

– Not as above.... 5

5. Tarsal formula appears to be 5-5-5 10

– Tarsal formula less or appear to be less than 5-5-5.. 6

6. Tarsal formula 5-5-4... 7

– Tarsal formula not 5-5-4.. 8

7. Each tarsal claw split to appear double. ....(blister beetles) ***Meloidae***

– Tarsal claw not split. ....(darling beetles) ***Tenebrionidae***

8. Tarsal formula appears to 3-3-3 (actually 4-4- 4) .... (ladybird beetles) ***Coccinelidae***

– Tarsal formula appears to 4-4-4 (actually 5-5-5)...... 9

9. Eyes notched, antennae longer, often longer than body ....(long horned beetles) ***Chrysomelidae***

10. Hind trochanter with greatly enlarged lobes. 11

– Hind trochanter without enlarged lobes.. 12

11. Clypeus wider than distance between antennae sockets, head usually broader than prothorax ....(tiger beetles) ***Cicindelidae***

– Clypeus narrower than distance between antennal sockets, head usually narrower than prothorax... ....(ground beetles) ***Carabidae***

12. Elytra short, 3 or more tergal exposed. 13

– Elytra covering the abdomen.. 14

13. Three tergal exposed.... ....(carrion beetles) ***Silphidae***

– Five or more tergal exposed... ....(rove beetles) ***Staphylinidae***

14. Pronotum enlarged and with lateral flanges, head difficult to see from dorsal view.. ....(fireflies) ***Lampyridae***

– Pronotum without such enlarged flanges.. 15

15. Pronotum with sharp posterior angles at lateral corners... ....(click beetles) ***Elateridae***

– Pronotum without sharp lateral angles.. 16

16. Antennae elbowed, mandibles greatly enlarged in males ....(stag beetles) *Lucanidae*

– Antennae not elbowed.. 17

17. Antennae highly lamellate.. ....(scrab beetles) ***Scarabaeidae***

– Antennae not lamellate or poorly so.... 18

18. Large black shiny beetles, often with "horn" on head ....(passalid beetles) ***Passalidae***

– Not as above.... 19

19. Rounded body with long legs, legs with yellow hair .... (spider beetles) ***Ptinidae***

– Not as above..... 20

20. Ventral surface with dense white hair..... ....*(dermestid beetles)* ***Dermestidae***

– Little if any ventral hair... 21

21. Elytra poorly sclerotised and flexible, elytra usually yellow or orange with black makings.. ....(soldier beetles) ***Cantharidae***

– Elytra strongly sclerotised and hard.... ....(metallic wood borers) ***Buprestidae***

(For details see Appendix III)

KEY TO FAMILIES OF ORDER ISOPTERA

1. Tarsi five segmented; hind wings with anal lobe.. ....***Mastotermitidae***

– Tarsi four segmented; hind wing without anal lobe. 2

2. Frontonel absent in all castes; empodium present (except on *Protermes*)... 3

– Frontonel present in all castes; empodium absent. 4

3. Ocelli absent, pronotum not always broader than head in soldier... .... ***Hodotermitidae***

– Ocelli present; pronotum broader than head.. ....***Kalotermitidae***

4. Scale or stump of forewing large, distinctly larger than that of hind wing; wing membrane often reticulated; worker with flat pronotum (except in *Rhinotermes*)... ....***Rhinotermitidae***

– Scale of forewings small; wing membrane never strongly reticulated; wings hairy; worker and soldier with a saddle-shaped pronotum.... .... ***Termitidae***

Simplified key to Order Orthoptera

1. Pronotum extending backwards almost to the tip or abdomen or beyond tip of abdomen. ....***Tetrigidae***
2. Front tibia and femora greatly enlarged for digging with large black tooth like process.. ....***Gryllotalpidae***
– Front tibia and femora not greatly enlarged for digging. 3
3. Hind femora much longer than thicker than middle femora.. 4
– Hind femora about the some length and thickness as middle femora. 7
4. Tarsi 3 segmented.. 5
– Tarsi 4 segmented.. 6
5. Antennae about as long as pronotum... ....***Acrididae***
– Antennae much longer than pronotum.. ....***Gryllidae***
6. Wings absent or vestigial, mesonotum and metanotum similar to pronotum, prosternal spines absent. ....***Gryllacrididae***
– Wings large, mesonotum and metanotum very different from pronotum, prosternal spines usually present..... .... ***Tettigonidae***

KEY TO FAMILIES OF BUTTER FLIES

1. Fore wings with vein 7 forked. Antennae approximate at base................ .2
– Fore wings with all veins arising from the cells. Antennae wide apart at base, and often with a hooked club.. ....***Hesperidae***
2. Hind wing with a precostal vein (except in a few Pieridae).. 3
– Hind wing without precostal vein..... ....***Lycaenidae***
3. Fore legs fully developed in both sexes... 4
– Fore legs not perfect in one or both sexes. 5
4. Hind wing with veins 1 b absent.. ....***Papilionidae***
– Hind wing with vein 1b present... ....***Pieridae***
5. Forelegs imperfect in both sexes.... 6
– Fore legs imperfect and brush like in the males, developed for walking in the females ....***Riodinidae***
6. Both wings with the cells closed, cross veins present. 7
– Hind wing with the cells open, cross veins absent. 9
7. Forewings with veins 1a forked with 1b.. ....***Danaidae***

– Fore wings with veins 1a free.. 8

8. Fore wings with one or more veins inflated or swollen at base, wings normally scaled, not spotted.. ....***Satyridae***

– Fore wings without any vein inflatted at base, wings thinly scaled and spotted. ....***Acraidae***

9. Palpi small, narrow and sharp infront of.. ....***Amathusiidae***

– Palpi large, broad, rounded infront cell of wing usually open. ....***Nymphalidae***

KEY TO LEPIDOPTERA (Important families only)

1. Antennae clubbed...**(Butterflies)** 2

– Antennae not clubbed. ....**(Moths)** 6

2. Terminal hook usually projecting from clubbed antennae ....**(Butterflies)** 3

– Terminal hook lacking...(Skippers) ***Hesperiidae***

3. Prothoracic legs much reduced.......................................(Brush-footed butterflies) ***Nymphalidae***

– Prothoracic legs normal sized.... 4

4. Large wingspread, usually 2 inch, hindwings with tails. ....(Swallowtails) ***Papilionidae***

– Not as above.. 5

5. Margin of eye notched where antenna located ....*(blue coppers)* ***Lycaenidae***

– Margin of eye without notch. ....(white and sulphurs) ***Pieridae***

6. Eye spots on wings, wingspread over 3 in... ....(giant silkworm moths) ***Saturnidae***

– Eye spots usually absent... 7

7. Antennae pectinate but not up to tips ....(royal moths) ***Citheroniidae***

– Antennae either pectinate up to tips or not conspicuously plumose. 8

8. Hindwings reduced, fits in indented forewings. ....(sphinx moth) ***Sphingidae***

– Hindwings normal, no indentation in forewings 9

9. Abdomen coloured other than brown, forewings either white or brightly striped.. ....(tiger moths) ***Arctiidae***

– Abdomen light yellow-brown to dark brown. 10

10. Body large when compared to wing-size, no large palpi extending beak like infront of head 11

– Body comparatively small, large beak-like palpi extending forward of head.. ....(pyralid moth) ***Pyralidae***

11. Antennae distinctly pectinate ... .(tent caterpillar moths) ***Lasicampidae***

– Antennae not distinctly pectinate... ....(noctuid moths) ***Noctuidae***

KEY TO SIPHONAPTERA (APHANIPTERA)

1. Head with transverse suture above the antennae.. 2

– Head without such a suture. 4

2. Head on each side with sclerotised flaps projecting downwards, parasitic on bats... ....***Ischnopsyllidae***

– Head without such flaps.. 3

3. Occipital region with a transverse dorsal thickening, ctenidia (comb of spines on gena or genial comb) present... ....***Macropsyllidae***

– Occipital region not as above, no ctenidia present.. ....***Hystrichopsylldae***

4. Labial palpi annulated... 5

– Labial palpi not annulated.. 8

5. Eyes usually present, few setae in a single row on abdominal segments.. ....***Pulicidae***

– Eyes usually absent, numerous setae in double row on abdominal segments 6

6. Genal ctenidia absent.. 7

– Genal ctenidia present.. ....***Ctenophthalmidae***

7. Head with a notch or tubercle in fronts. ....***Dolichopsyllidae***

– Head without a notch or tubercle.. ....***Uropsyllidae***

8. Maxillae with long curved lamina, head rounded... ....***Hectopsyllidae***

– Maxillae without or with small lamina, head angulared... .... *Tungidae*

KEY TO DIPTERA (Important family and Sub orders)

1. Antennae filiform to plumose with atleast 7 segments .... ***S.O. Nematocera***, 2

– Antennae with fewer than 7 segments. 4

2. Mesonotum with V-shaped suture... ....(crane flies) ***Tipulidae***

– Mesonotum without V-shaped suture.. 3

3. With long beak, scales on wing veings.. ....(mosquitoes) ***Culicidae***

– Beak and scales on wings absent... ....(midges) ***Chironmidae***

4. Arista absent.... ....***S.O. Brachycera***, 5

– Arista present.... .... ***S.O. Cyclorrhapha*** 7

5. Vertex indented between eyes... ....(robber flies) ***Asilidae***

– Vertex not indented.... 6

6. Body with dense hair, head not broad from above ....(bee flies) ***Bombylidae***

– Body without dense hairs, head broad from above. ....*horse and deer flies)* ***Tabanidae***

7. Spurious vein present in wing... ....(flower flies) ***Syrphidae***

– Spurious vein absent in wing. 3

8. Ridge or distinct bulge beneath and behind scutellum, arista bare... ....(tachinid flies) ***Syrphidae***

– No distinct ridge or bulge, arista atleast partly plumose.. 9

9. Hypopleuron without row of bristles... ....(muscid flied) ***Muscidae***

– Hypopleuron with distinct row of bristles.. 10

10. Notopleuron with 2 large bristles, arista plu mose to tip.... .... (blow flies) ***Calliphoridae***

– Notopleuron with 4 large bristles, arista plumose only in basal half.... ....(flesh flies) ***Sarcophagidae***

CHECKERBOARD KEY OF SUPERFAMILIES OF ORDER COLEOPTERA

Superfamily	Larval types	Habitat	Food habit	Visible abdominal sternites	Corneous abdonimal tergites	Antennae	Elytra	Tarsal Formula	Miscellaneous Structures
Staphyli-noidea	Campodei-form	terrestrial	Scaveanger & predator	5-8	3 or more	Simple or clavate	Often abbreviated	3-3-3 to 5-5-5	wings without crossveins.
Hydrophi-loidea	Campodei-form	mostly aquatic	Scavanger	5 rarely 4 or 7	not more than 2	Clavate 6-9 segmented	covering most of abdomen	5-5-5	Maxillary palpi often longer than antenae
Elateroidea	Campodei-form	terrestrial	Phytophagous root feeder wood borer	5	not more than 2	usually serrate or pectinate	Usually covering most of abdomen	5-5-5	Prosternal process fitting into mesosternum
Cantharoida	Campode-iform	terrestrial	mostly predator	5-8	not more than 2	Filiform to clavate	Rarely abbreviated	5-5-5 or apparently 4-4-4	Integument soft
Dryopoidea	Variable	aquatic	Phytophagus	usually 5	not more than 2	Variable often serrate	Usually covering mostly abdomen	5-5-5	prosternal process fitting into mesosternum
Dascilloidea	Variable Campode-iform to eruciform	aquatic & terrestrial	Phytophagus or scavanger	5	not more than 2	variable	Rarely abbreviated	5-5-5	head not constricted posteriorly
Cucujoidea	Variable frequently campodeiform	terrestrial	Phytophagus or predaceous	5	not more than 2	variable but usually clavate	Covering most of abdomen	3-3-3 to 5-5-4 or 5-5-5	head not constricted posteriorly
Mordelloidea	Variable	terrestrial	Phytophagous	5-6	not more	Variable	Covering most	5-5-4	head strongly

Superfamily	Larval types	Habitat	Food habit	Visible abdominal sternites	Corneous abdonimal tergites	Antennae	Elytra	Tarsal Formula	Miscellaneous Structures
	Hypermeta-morphosis common		parasitic		than 2		of abdomen		constricted posteriorly
Tenebrio-noidea	Canpodeiform hard integument	terrestrial	mostly scavanger	5	note more than 2	Variable but never pectinate	Covering most of abdomen	5-5-4	head not constricted posteriorly
Ptinoidea	cruci form leg short	terrestrial	phytophagous mostly wood borers	5	not more than 2	clavate usually	Covering most of abdomen	5-5-5	head often retractable
Scaraba-eoidea	Cruci form leg short	terrestrial	phytophagous or scavanger	6 rarely 5	hot more than 2	lamellate	Covering most of abdomen	5-5-5	legs usually fossorial
Ceram-bycoidea	cruci form legs short or absent	terrestrial	Phytophagous leaf feeders & wood borers	5 rarely 6	not more than 2	filiform rarely serrate	rarely abbreviated	apparently 4-4-4	third tarsal segment usually bilobed

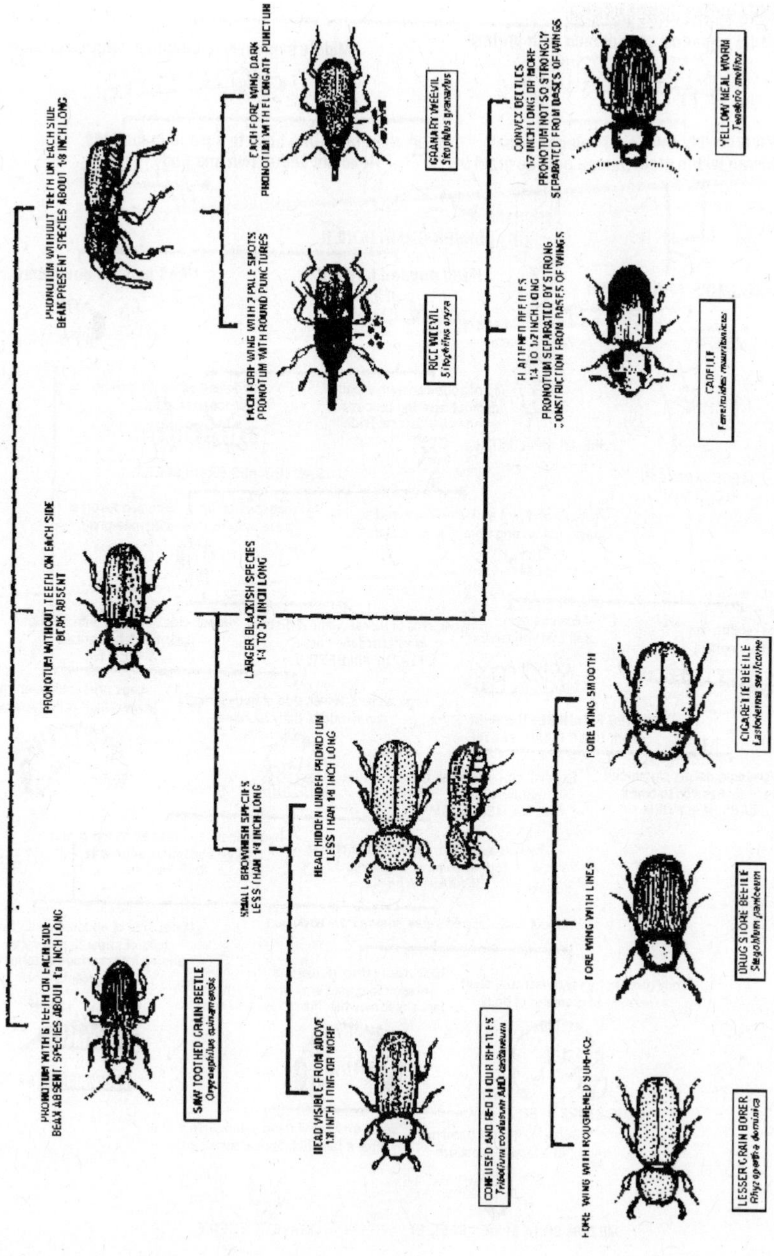

Fig. 4.1. Pictorial key to some common beetles associated with stored foods. (Courtesy: U.S. Department of Health, Education, and Welfare).

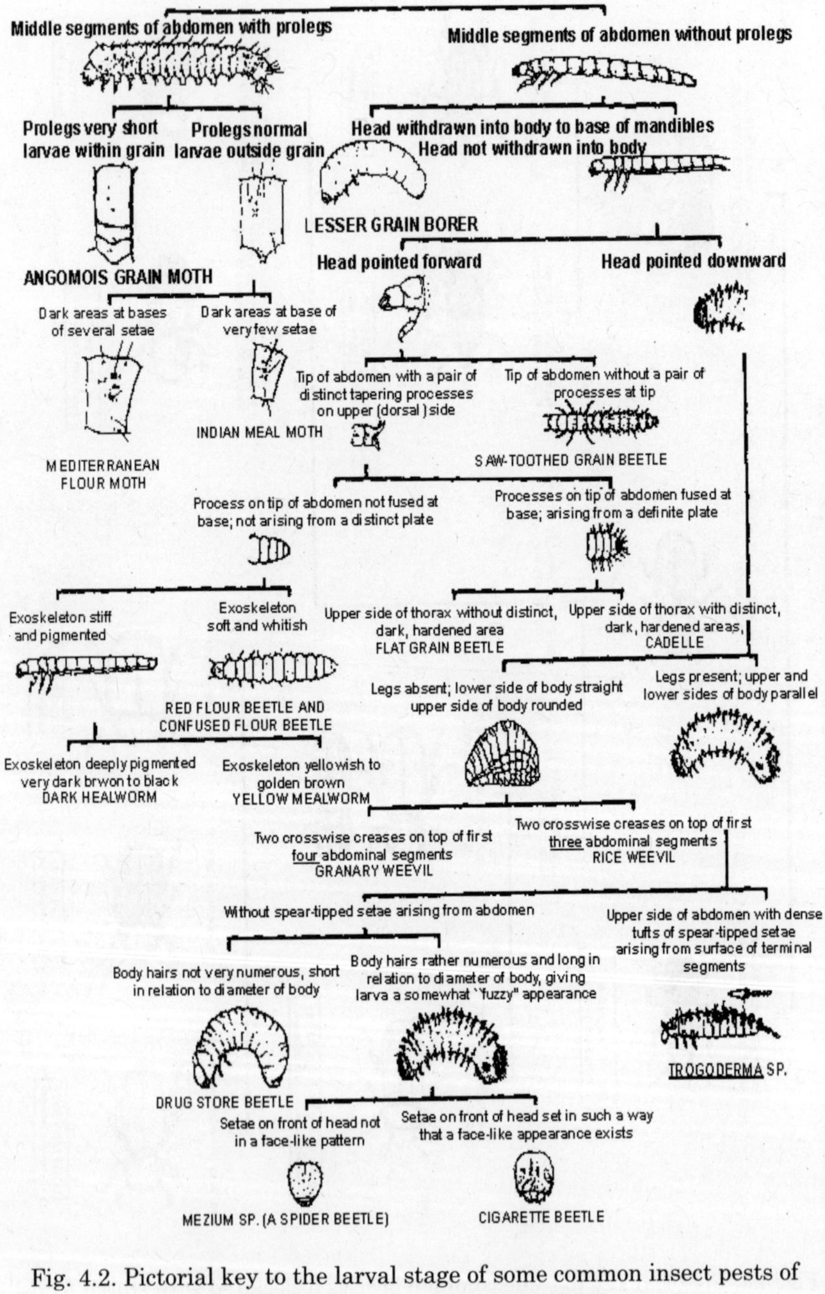

Fig. 4.2. Pictorial key to the larval stage of some common insect pests of stored grain and other food products.

CHAPTER - 5

BRIEF NATURAL HISTORY OF SOME COMMON INSECTS

1. Earwigs (Forficulidae)

Earwigs are medium sized insects. The forceps at the extremity of the abdomen is characteristic of the family. They are adapted for running swiftly on the surface of the soil. Round white eggs are laid in a mass in the ground or in shelter, the female in some cases guard them until they hatch. *Parental care* is found. The youngs are white at first and resembles with Thysanura. The Earwigs are most active in the rains and damp weather. In irrigated land they are active throughout the year.

Collecting : They can be collected from under stones, among decaying vegetation and fallen leaves among debris. Some are found in houses especially in damp places such a bathrooms in the hot weather, at roots of the plants in cold weather. A few come to light. **They are pinned through the right wingcase**.

2. Praying Mantis (Mantidae)

The family is recognised by the *raptorial forelegs* in which the tibia works in opposition to the femur like the blades of a scissors and both are wholly or partially spined. The colour and form is cryptic that resembles the natural objects. The females deposits her eggs in a characteristic large egg case called *ootheca*. It is fixed to a plant. The egg case is made of gummy matter secreted by the female which comes out as a frothy mass and sets hard in a short-time. The female assumes firm position on the plant and extrudes a mass of frothy gum and with the end of abdomen shapes it. As soon as the base is formed and some amount of gum used, eggs are laid in the midst of the gum. The emission of eggs and gum continues, the eggs in the middle, the gum round, until the whole egg mass is built up, layer by layer, when she finishes it off with gum and the whole hardens to a water tight object firmly secured to the plant. The eggs are in regular rows inside the egg case and the whole mass will last through the winter on the plant. The young mantids emerge which resembles an ant. Nymphs exhibits *mimicry*. They are active and leads an active life till they are full grown. The eggs are laid is November and hatch in March. They are abundant in jungles but still commonly in the cultivated plains.

Collecting : They are found upon bushes, in grass, on the bark of trees. Sometimes come to the light also. They are best **pinned through the right wing case or prothorax**, the left-wing being set.

3. Painted grasshopper (Acrididae)

The host plant in Aak. It is also found on other milk weeds. It is brightly coloured in blue and yellow, living openly on its food plants. There are atleast two broods in a year. The last laying passes the winters. The nymph is coloured yellow with black stipples and red spot. It is of bad taste and a white substance is exudates on disturbing which is of characteristic smell. The males are smaller then females. Members of the family acrididae are herbivorous and most injurious to agriculture.

Collecting : They can be easily collected by hand or net from the vegetation. Few comes to light. **They are pinned through the right wing or thorax**. The left wing should be spread.

4. Ant lions (Myrmeleoninae)

They are recognised by short clubbed antennae, the wings are usually large and of equal size often marked with brown and black. The eggs are laid in sand or earth. The larva is flat. Jaws are predominant in the larva projecting in front of the head. It has free wandering life or live in pits in the sand. The larva prepares these pits. It commences by going round in a circle, moving backward, its body making a furrow in the soil. It throws out sand with the help of broad head and excavate a round pit. It then lies there in the pit motionless and wait for some insects that walks over the pit. The insect fall in the pit and comes in the reach of the jaws. The insect is seized and sucked out and the dried shell is thrown out. Ant are the large part of its diet. The adults are found from March to November. The pupation takes place in a cocoon in the same in sand near the pit. Silken cocoon is produced from the apex of the abdomen. Imago flies clumsily. It is nocturnal but seen flying during the day also in long grass. An unpleasant odour is diffused from their body. They are attracted to the light.

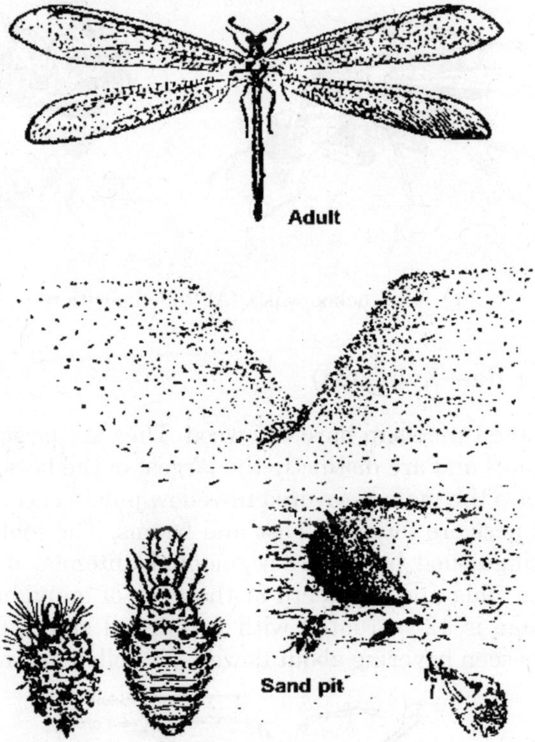

Fig. 5.1. Ant lion in pit, b. Larva

Collecting : They are attracted to light. No special method of collecting is required.

5. Cuckoo wasp (Chrysidae)

They are small insects with dusky wings and bright metallic green or blue green colouring. Abdomen is with three visible segments. Female has a long retractile ovipositor. The body structure is hard designed to enable the insect to curl into a ball. All are parasitic upon aculeate Hymenoptera.

Collecting : Cuckoo wasps are common in houses and in the open they require to be caught with a net. Pinning is done like other Hymenopterans.

Fig. 5.2. Cuckoo wasp. (A) Rolled position.

6. Carpenter Bee (Xylocopa)

They make tunnels in hard dry wood. They are large black insects with dark wings and are distinctly the largest of the bees in the plains. In some species the male is covered in yellow pubescence, the female in black. Their nest are found in post and beams. The male of *Xylocopa* may be distinguished by the 13-segmented antenna, a single apical spine on hind tibia and two teeth at the apex of mandibles. In female these antennae is 12-segrented with two apical spines on hind tibia. The insects is seen hovering about flowers for pollen and nectar.

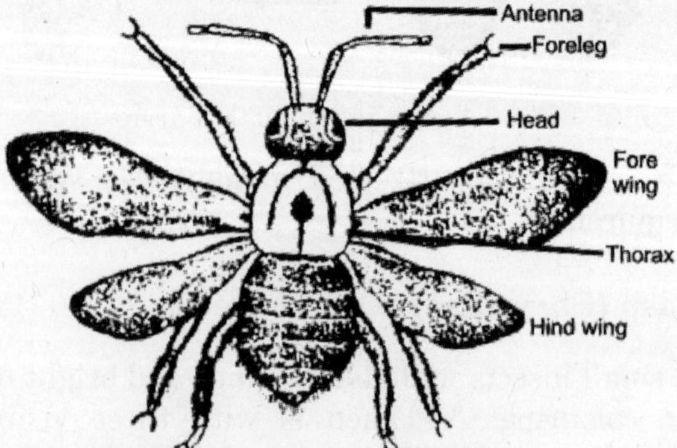

Fig. 5.3. Carpenter bee (Xylocopa)

Photo: Natural habitat of Xylocopa.

7. Dung Rollers (Scarabaeidae)

The members of this family exercise a very important function in the economy of nature. They clean the surface of the earth of the excrementitious matter deposited on it and carry it to the soil thus provide valuable manure. The Beetles collect in dung, feeding upon it and making it into balls which they roll over the surface of the ground and take into the soil where either they feed on it or use it as food for their future youngs. The ball of the dung is divided into pieces in which an egg is laid. The larva eats the dung. The flat head of the adults is used as shovel for digging, forelegs helps in excavating hole in the ground. In hot and dry weather the dung attracts the adult beetles. They cut and shape it into a ball and roll it. The sense of smell of these beetles is very acute. The larva is never seen and live below ground. *Helicopris* is commonly found in the plains.

Collecting : They are seen abundantly from April to June but also common in rainy season. **They are pinned through the right of elytra.**

Fig. 5.4. Dung Rollers in their natural habitat.

8. Click Beetles (Elateridae)

Click beetles generally come to the light during rains. They are flat and stout. When the beetle falls upside down on its back, it does not struggle helplessly. It arches up its body on the tips of its head and abdomen, then suddenly draws the spine out of its pit and jerks itself up, producing a sudden clicking sound and then lands right side up on its legs. The larvae are cylindrical and elongated that feed on the root of plants.

The prothorax of the adult is large and powerful, fitting loosely but accurately to the elytra. The lateral angles prolonged backwards, on the vental surface is a process which passes into mesothorax in which is a cavity fitted to it Development is slow. Several years are occupied in metamorphosis.

Collecting : They are common everywhere. They can be best collected in rainy season. **They are pinned** (as other coleopterans) **in right side of the elytra.**

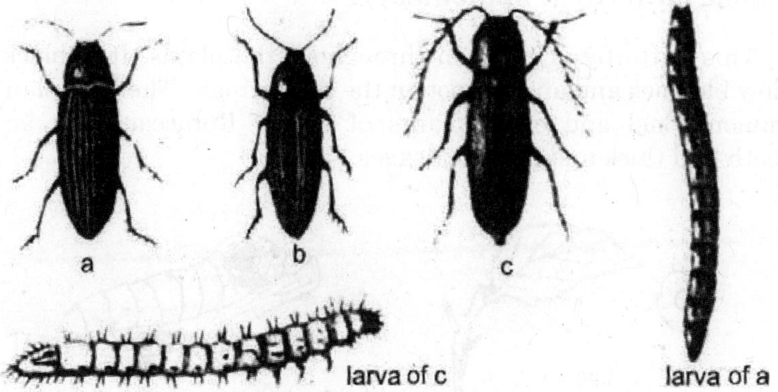

Fig. 5.5. Click Beetle
(a - Lined click beetle, b - scarlet click beetle, c- comb horned click beetle)

9. Plain tiger (Danaidae)

Brightly coloured butterflies with fore legs clubbed. The caterpillar of plain tiger is grey coloured with five yellow and black bands on each body segment along with a yellow stripe. It bears three fleshy processes. It feeds on *Calotropis* leaves. *Danaus genutia* is somewhat larger and darker butterfly and its wing veins are broadly blackened. The pupa is beautiful green and shiny with golden spots on anterior end and a golden bordered line round the posterior end. The food plant is aak. It is common throughout the year but abundant in October-November.

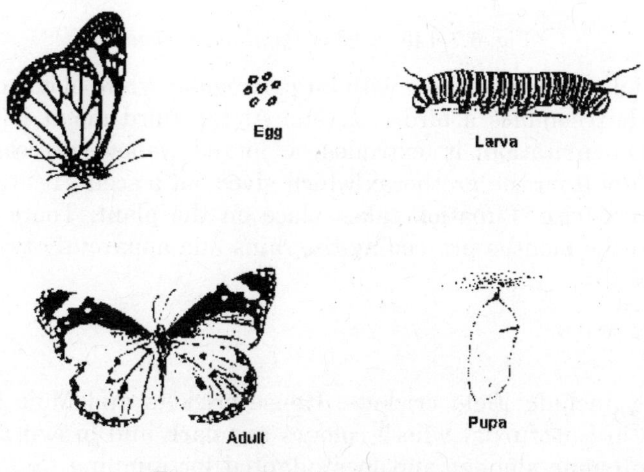

Fig. 5.6. Life cycle of Plain tiger.

10. Lime butterfly (Papilionidae)

This butterfly is common throughout the plains. It is black with yellow blotches and an eye spot on the hind wings. The host-plants are Lemmon, Bael and other plants of family Rutaceae. The larva is smooth and thick and feeds on leaves.

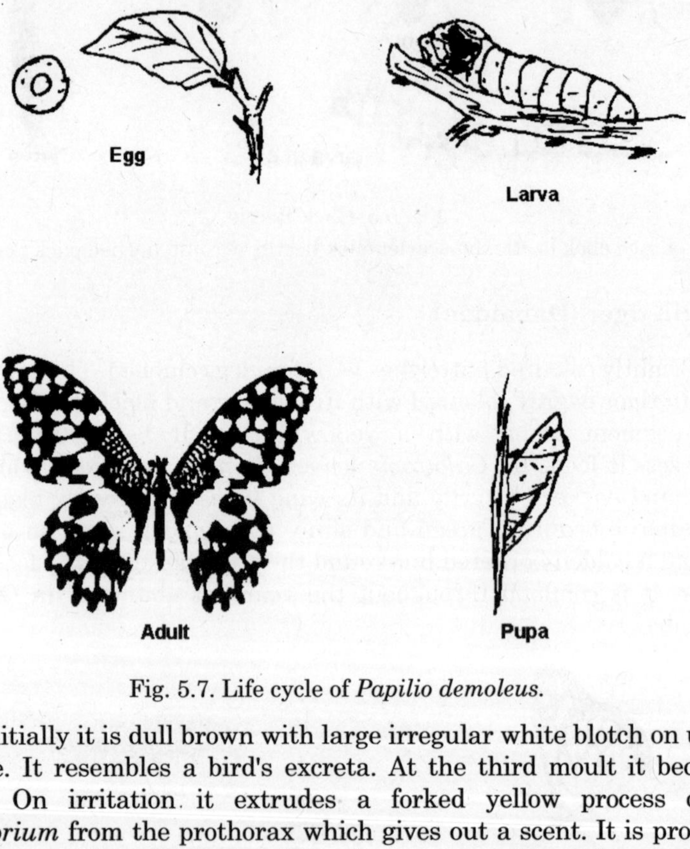

Fig. 5.7. Life cycle of *Papilio demoleus.*

Initially it is dull brown with large irregular white blotch on upper surface. It resembles a bird's excreta. At the third moult it becomes green. On irritation it extrudes a forked yellow process called *osmatorium* from the prothorax which gives out a scent. It is probably protective device. Pupation takes place on the plant. There are two broods in the months preceeding the rains and apparently two during these months.

Cricket

They include Field crickets, House crickets and Mole crickets. Crickets are nocturnal whose colours are dark and brown and grey tints. Antennae slender and legs adapted for jumping (*Saltatorial*).

The elytra of males have a large musical apparatus and females possess long ovipositor. The musical sound has given the name 'cricket'. All of us have heared it by night in the fields and in houses. They live solitary except when they are disposed to the chirping society or group. They find shelters in crevices, holes cracks etc. The house cricket is yellow grey and more or less marked with brown tints. It seeks its food from the warm places like kitchens. In towns it prefers bakeries and the neighbourhood of ovens. It is most commonly found household pest in India.

Field cricket

Field crickets are black in colour and nocturnal in habit. It is of great annoyance even in the gardens, lawns in houses.

Fig. 5.8. Field Cricket and its habitat

11. Mole cricket

Mole cricket (*Grylotalpa*) is wonderfully adapted for fossorial life. It has fossorial fore-legs adapted for digging the ground. Gardens and vine yards are most preferred dwelling places. They make gallaries destroying everything which comes in their way. They cut through the roots and eat fine underground twigs. It principally feeds on vegetable matter but also takes worms and other smaller insects which comes their way. Their excavations consist of vertical shaft more or less deep and long horizontal gallaries. Mole crickets are regular miner. Female lay eggs is the deepest part of the mine. They destroy roots of garden plants and other vegetation of their habitats. For most of the times, they live under ground and rarely leave their holes. They often come to the light in houses in rainy season when their mines are flooded. The female lays eggs in under ground chambers. She picks them up one by one and lick to clean, feels them with antennae (feelers) and tenderly place them back. This solicitude continues till they hatch, thus they reveal organised parental care.

Fig. 5.9. Grylotalpa in its natural habitat

12. *Water Beetle* (Hydrophilidae)

They are adapted to live in water. The legs are *natatorial.* The females have some glands in the abdomen which produce a silky substance that covers their eggs with an impervious cocoon.

The female crawls over the aquatic plant with its head downward. It lays its eggs and immediately drags its silk drawers over and around them and it envelops them in a cocoon. The cocoon is fixed either to a leaf or to a stem. The colour of the cocoon is light grey and short oval in shape with a long conical pedicel.

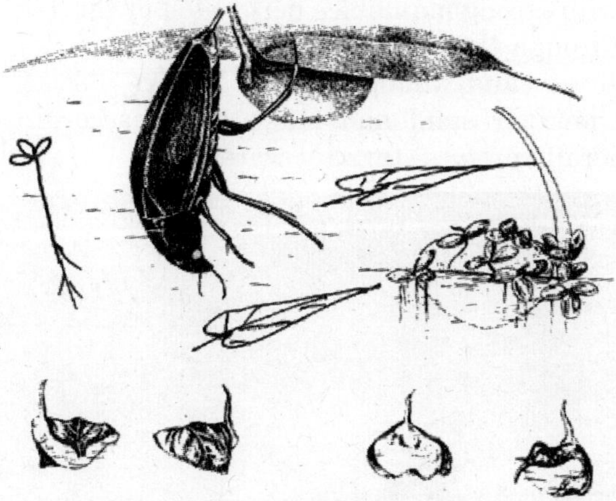

Fig. 5.10. Egg cocoon of *Hydrophilus*

When the young larvae hatched they begin to feed and grows very rapidly.

The genus *Hydrophilus* contains large insects of an oval form having antennae with nine joints, the last four forming the club. Sharp pointed spurs are capable of inflicting injuries (wounds) if handled incautiously. The adults are olive black in colour and swims with great facility. The hind legs act as oars. It is never horizontal at the surface of the water, but it pokes out its head above the water level, folds its antennae, the club shaped ends of which are partly hollowed. It carries downwards a small bubble of air, which is pressed against the sides of the thorax and of the abdomen that comes in contact with respiration system.

Dytiscus (Dytiscidae)

This beetle is also adapted for aquatic and sub aerial life. They are like land beetles therefore, obliged to come to the surface every now and then to breathe air. The spiracles are situated on the upper part of the body and are protected by the elytra which fit the upper and sides of the body so that no water can get underneath them. During breathing, it lets itself float to the top of the water. The upper part of the back projects above the surface. The elytra are raised up, expiration takes place and supply of fresh air is inspired. Afterwards the elytra shuts down that collect some air which is used from time to time.

Dytiscus is carnivorous. Their mandibles are strong. Their larvae are aquatic with strong mandibles perforated at the extremity so that it can suck through them. The larvae constantly chasing away other insects, molluscs and young fishes. Mature larvae undergo for pupation in mud or sand and make a cell according to its size. Dytiscus generally prefers stagnant waters.

Fig. 5.11. Natural habitat of Dytiscus

13. Whirligig Beetles (Gyrinidae)

Peculiar water beetles known for circling upon the water with rapidity. Due to this extra ordinary movement, their name *whirligigs* (Tourniquets, French word). They are small and their body is oval, more or less convex, legs adapted for swimming, the middle and hind ones act as broad oars and the front legs as rudders. They are generally of brilliant metallic bronze colour and looks like dancing pearls on sun shining over them. Nature has given them four eyes. Their compound eyes are divided into an upper half for vision in the air and a lower half for vision under the water. When caught, they exudate a milky fluid of an extremely strong disagreeable odour which remains for a considerable time afterward.

The female gyrinid beetle lays eggs upon the surface of the leaves. The larvae are long thin and white when full grown are very active and voracious. The last segment has four movable hooks by which it jumps. They get out of the danger by rapid jumping. On metamorphosis, they leave water and take up abode on leaves of the aquatic plants and spin coccoon.

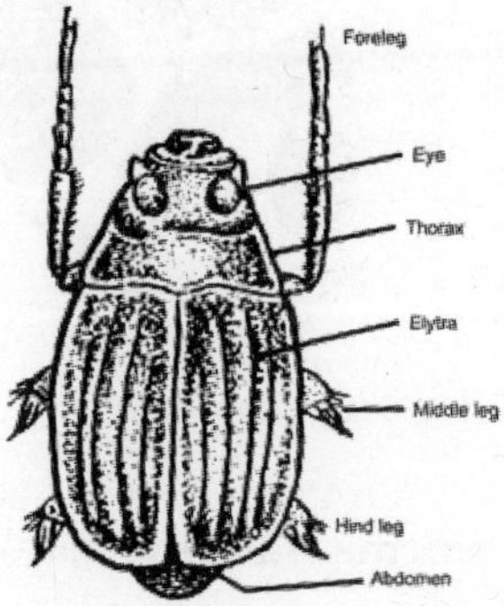

Fig. 5.12. Gyrinid beetle (*Dineutes indicus)*

14. Lady Bird Beetle (Coccinellidae)

They are very attractive and pretty. They are round, oval red or brown and black spotted. Larvae are bright yellow or grey, three little spots on the sides besides black spots and little bunch of spines. The seven spotted Lady Bird beetle, (*Coccinella septempunctata*) is widely distributed in India. Both adults and larvae feeds upon aphids of various species. They are potent agent of Biological control of aphids. *Chilomenes* is another Lady Bird Beetle of India, small reddish or yellow beetle with six black wavy stripes across the body. A single beetle has been reported to devoure as much as 200 aphids in a day.

Fig. 5.13. Seven spot lady bird beetle

15. Fire flies or Glow worms (Lampyridae)

The glow worm is the wingless worm like adult female or the larvae. The fire fly is winged male. They are common is the plains during the wet season. The curious phosphorescent light comes from under parts of the abdomen near the tip below 8th abdominal segment. This oval patch is the luminous organ which emits light at the will of the insect. It is nocturnal and carnivorous. It feeds on soft-snails, slugs and other mollusca. It seize the snail by its legs, curls over on it back and eat bit by bit. The light produced by the glow worm is greenish white, pale and cold.

Fig. 5.14. Glow worm.

16. Long horn Beetles (Longicorn)

Family - *Cerambycidae*

They are very strong and stout with enormous mandibles differently toothed according to species. According to their peculiar habits, its lower lip is provided with modifications in form. They are leaf eaters. The diversity in the structures of mouth parts depends upon the nature of the vegetable matter they feed. They are cosmoplitan. The larval segments are little swollen, the first is largest and is covered by means of a leathery plate. The larvae lives in the tunnels of the trees. They never come to light therefore colourless. As they feed the sapwood they form galleries, they have strong jaws and head. They do not walk much in the galleries, they lac legs therefore apodous. The larvae make cocoon of fragments of wood and little bit of vegetable tissue with their saliva. The larvae of *Acanthophorus serraticornis* lives under the bark of large trees. *Acanthophorus* is large beetle of brown red (coffee colour) elytra. Antennae long about 2/3 of the body, tarsi 5 segmented and tibia with two spurs. The mandibles are massive and used to make galleries in the tree trunk. Many cerambycids are pest of large trees like mango.

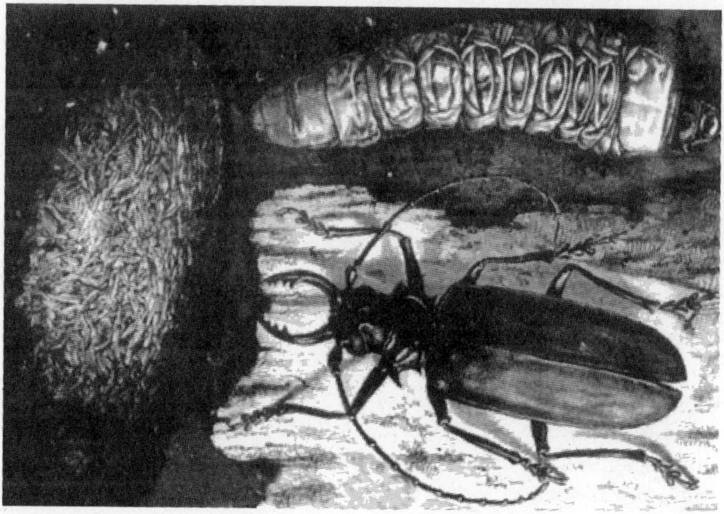

Fig. 5.15. Long horn beetle

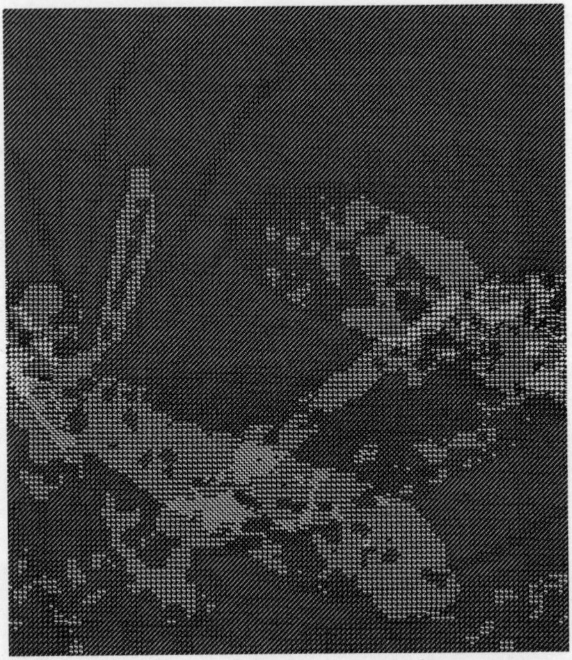

Fig. 5.16. Natural habitat of Caddis Fly

17. Caddis fly

They undergo complete metamorphosis. Their wings have no cross veins and are peculiarized in the presence of small hairs. The adults fly in the evening near by rivers streams and ponds. The females deposit eggs in water and covers them in glutinous mass to the stones or aquatic plants. The larva of caddis fly (order : Trichoptera) is caddis worm. It construct silken thread lined cabin of loose stics, bit of leaves, pebbles plant fibres etc. The larva has 3 pair of legs and filamentous delicate gills for breathing. The tail has powerful hook to adhere itself in the cabin. It with draws into the cabin on slight disturbance. In running water its case is made up by pebbles so that it is protected against water waves. It feeds on algae and other aquatic plants. It also feeds on larvae of other insects.

18. Velvet ant (Mutilla)

The females are devoid of wings and can run on the ground like worker ants. They are solitary consist of male and females only. The females of *Mutilla* is black, its thorax red and has three grey bands on the first three segments of the body. They are found in woods, walking slowly on the ground and seen entering in the holes in the earth. The females have powerful stings and acrid venom together with strong and spiny legs.

Fig. 5.17. Mutilla in its Natural habitat

19. Flesh flies (Sarcophaga family - Sarcophagidae)

They devour dead bodies of all kinds. The leg less larvae called `fly blows' increase in size rapidly. Larvae *muscoid* mostly saprophagous but feeds on the flesh of animals also.

Some species cause *Myiasis* in man and other animals (cattles) species *Sarcophaga fasciata* deposit larvae in the open wounds of animals. Some species are parasitic on bees and wasp.

Fig. 5.18. Flesh flies (Sarcophaga and Calliphora)

Calliphora (Family: Calliphoridae)

Commonly called *Blue bottles* Large black or metallic green or blue, coppery or bronze coloured. Plumose arista and larvae muscoid. Mostly saprophagous or carnivorous *Calliphora* deposit its eggs on the soiled wool. The larvae attacks the flesh. Green bottle (*Lucilia*) cause cutaneous myiasis and infest sheep.

20. Horse flies or Gad flies (*Tabanus*)

It chase the animals in order to suck their blood (Bot flies chase for depositing eggs) they posses piercing and sucking mouthparts.

Fig. Life cycle of Tabanus

The eggs are laid on the skin, larvae bore into the skin, they moults twice and leave the host before pupation. Most species are blood sucking pest of horses and cattle. They often transmit *Surra* disease and *anthrax.*

CHAPTER - 6

APPLIED ENTOMOLOGY

(i) COMMON APPLIANCES USED IN INSECTICIDE APPLICATIONS

Appliances used in pesticide applications are broadly classified into (i) Sprayers (ii) Dusters and (iii) Granule applicators. Sprayers and Dusters are further categorised into (i) *Manually operated* and (ii) *Power operated.* Sprayers are used to disperse liquid formulations of pesticides whereas the dusters are used to disperse the powder form of pesticides.

1. SPRAYERS

They include **foot operated sprayers, rocker sprayers, compression sprayers, hand sprayers, knapsack sprayers** and **bucket sprayers.**

KNAPSAC SPRAYER

(1) It is the most convenient and efficient hand sprayer.

(2) It is provided with a tank which is generally flat to fit comfortably on the back of the operator.

(3) The pump is operated constantly to get a continuous discharge of liquid.

(4) A lever handle provided at the top or bottom of the tank operates the sprayer.

(5) Few strokes of the pump builds up pressure in the tank.

(6) The liquid is discharged into fine droplets-through a nozzle at the terminal end of delivery tube.

(7) It is suitable in small plots, hilly tracts and fruit gardens.

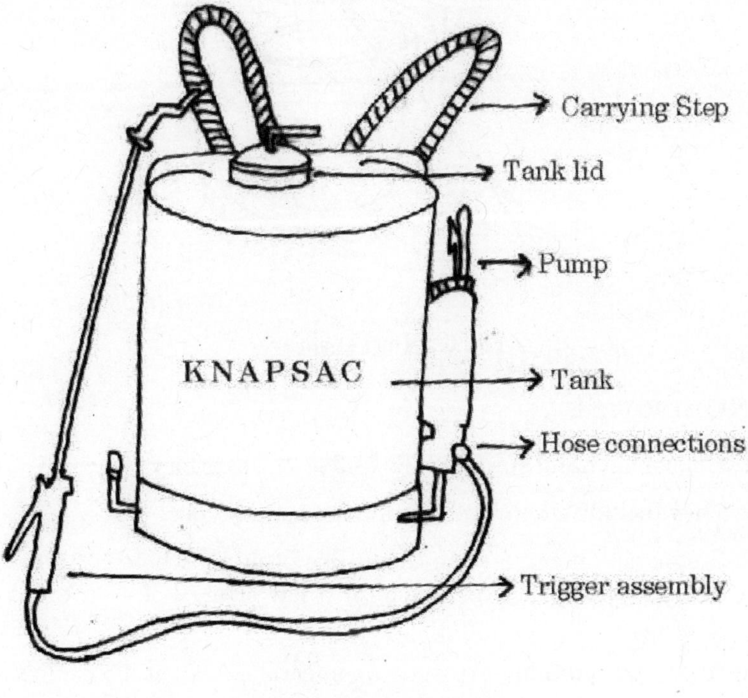

Fig. 6.1. Knapsac sprayer.

BUCKET SPRAYER

1. They are suitable for small scale spraying.
2. It is designed for use with a bucket containing the spray liquid.
3. It is provided with a foot rest for holding the pump in a proper position inside the bucket.
4. The up and down movement of the pump handle results in a continuous flow of liquid under pressure through the nozzle.
5. There are two types of bucket sprayers viz. *single barrel* (stirrup pump) and *double barrel*. Double barrel bucket sprayers can develop a pressure of about 60 pounds per square inch.

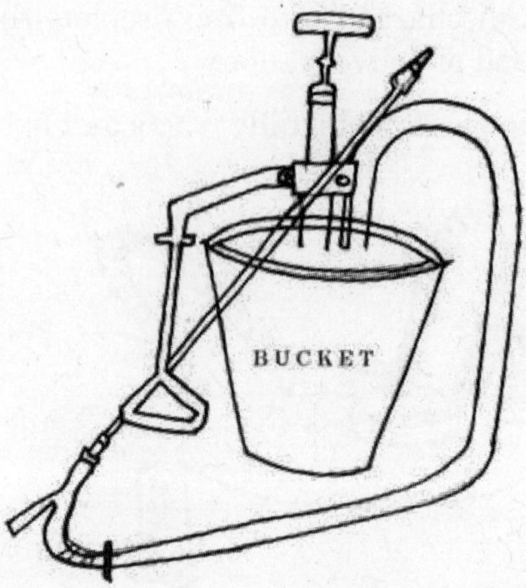

Fig. 6.2. Bucket sprayer.

HAND SPRAYER

1. Household hand sprayer falls under this category.
2. They include single and continuous action types.
3. Single action act only on the stroke of pump whereas continuous action type build pressure after a few strokes of the pump and gives a continuous spray.
4. They are suitable for spot disinfection against houseflies and mosquitoes.

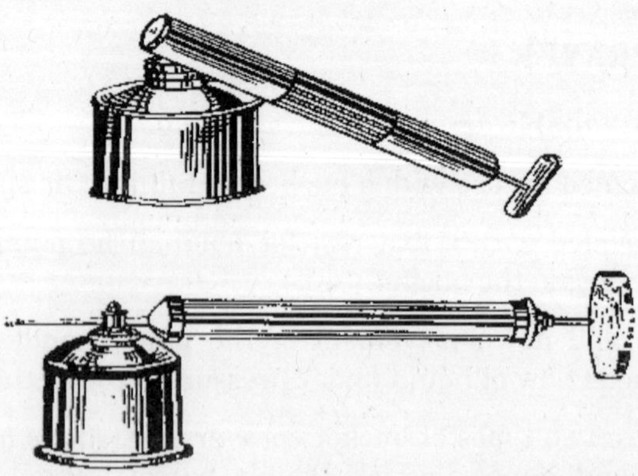

Fig. 6.3. Household hand sprayers. Single action (top) and continuous action (bottom)

2. DUSTERS

Hand operated dusters include *Rotary types, Plunger* and *Bellow type dusters*. Power operated dusters include Rotary type and Air jet machines. Given below is a description of *Rotary hand operated duster*.

1. This duster is designed to carried on the shoulder and belly of the operator.
2. It is provided with an agitator to stir the dust so that it will flow evenly through the discharge opening.
3. The main parts are *blower* with gear box, *hopper* and *hand crank*.
4. The blower sucks the dust from the hopper through the connecting pipe and discharges it with velocity.
5. The right hand works to turn the crank and left hand for handling the discharge tube.

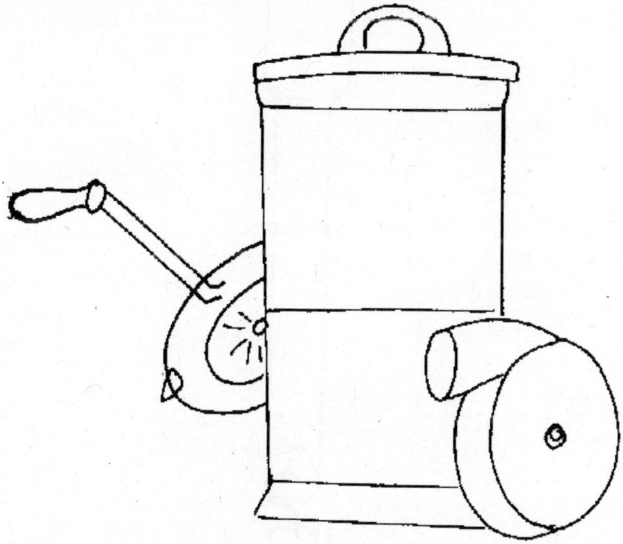

Fig. 6.4. Rotating Duster

PLUNGER DUSTER

1. This is used specially dusting plants in small areas without much waste of powder.
2. It is suitable for household dusting, kitchen garden and nurseries.

3. It is made up of air pump, a dust chamber and a discharge assembly consisting of straight tube with nozzle or deflector cap and the end of the dust chamber.

3. GRANULE APPLICATORS

1. They are designed to provide accurate and uniform distribution of granule of pesticides.
2. Most common categories are row crop granule applicator and trailor type applicator.
3. Row crop applicators are used to control pests on the plants or crops grown in rows whereas the trailor type is used in large farms.

Fig. 6.5. Plunger Duster

(ii) INSECT REARING

1. *Tribolium castaneum* (Herbst) (Red rust flower beetle)

(Coleoptera : Tenebrionidae)

Temperature - 30 ± 2°C

Humidity - 65 ± 5%

Food - Wheat flour + yeast

Procedure

Take a clean and sterilized wide mouth glass jar. Filled it half with sterilized wheat flour. Mix Brewer's yeast in a ratio of 100 : 5 (i.e 100 gm flour + 5 gm yeast). Obtain some adult *Tribolium castaneum* either from grocery shop where the grain are already infected with this pest. Release the adult beetles in sex ratio 1:1. Cover the culture jar with muslin cloth tightly held with rubber band. Keep the culture jars in the laboratory at 30 ± 2°C temp and 65 ± 5% RH. After 10-12 days the adults released earlier are removed by sieving and the eggs and larvae are allowed to develop till the generation of F_1 adult appears.

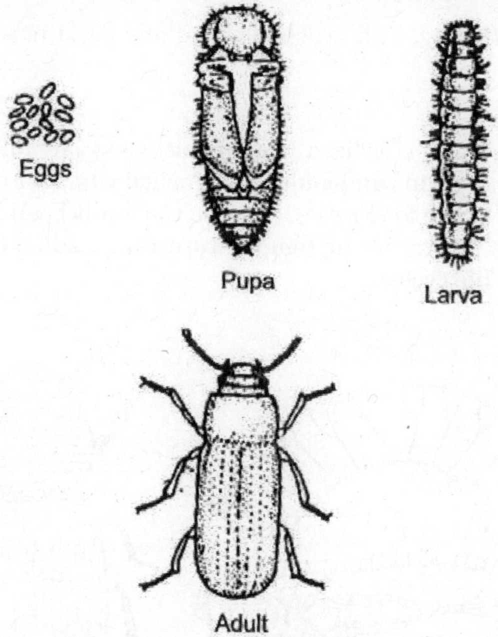

Fig. 6.6. Life history of *Tribolium castaneum*

The eggs are laid singly, delicate, white in colour and oval in shape. Incubation period lasts for 4-5 days. The newly hatched larva is an extremely delicate, translucent, white small slender and wormiform and gradually becomes creamish in colour. Its length varies from 6-8 mm. Full grown larva is cylindrical and passes through six instars. The total larval period varies from 22-30 days at temperature range 22.5°C-34.5°C and R.H. 70%. Pupal stage lasts for 1-2 days. The pupa is white, covered with bristles and gradually changes to light brown colour just before the emergence of adult. Mortality is maximum at 45°C. Adult Tribolium is *negatively geotropic*. Suji is the most preferred food (Sengupta, 1965).

Note : It is a severe pest of oilcake, copra, oilseeds, gram flour, suji, and a variety of condiments and dry fruits.

2. *Rhizopertha dominica* (Fabricius) (Lesser grain borer) (Coleoptera - Bostrichidae)

Temperature - 28 ± 2°C (26°C-28°C)

Humidity - 56-60%

Food - Wheat grains, crushed wheat and yeast in a ratio 100:1:1

Procedure

Take clean and sterilized wide mouth glass jar. Filled it half with sterilized wheat grains and add some crushed wheat or daliya also. Mix the yeast (1% of the food ratio). Release the adult beetles and keep the culture in the laboratory at temperature range 26-28°C and 60% RH and study the life cycle.

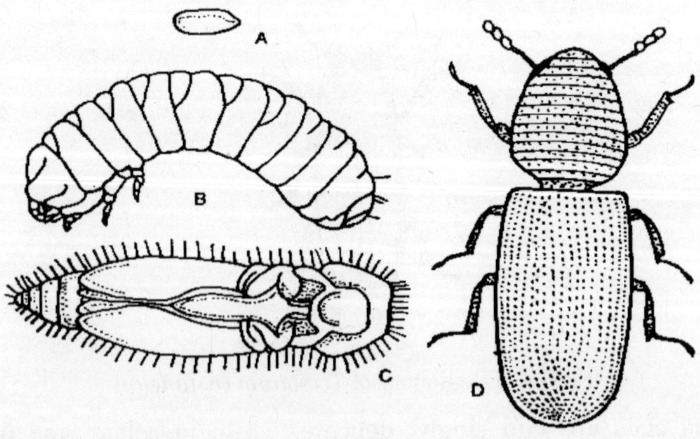

Fig. 6.7. *Rhizopertha dominica* Fab. - Life history (a) Egg (b) Larva (c) Pupa (d) Adult

Its life cycle was studied by Potter (1935). Adult female lays 300-500 eggs amongst the grains and hatching starts after 4-6 days. The larva is scarabae form and 0.3 mm. long and mature larva is recurved and incapable of locomotion. Larvae are white and slightly yellowish toward head.

Larvae are initially external feeders and the larval period lasts for 35-45 days. Pupation takes place on the surface of the grain and lasts for 7-8 days. Life cycle is completed in 6-8 weeks. Temperature and humidity conditions tends to influence the life cycle. In drier conditions it takes longer period.

Note: It is a pest of stored grain of all kind specially, wheat, barley, maize, rice, millets, sorghum, flour, dried potatoes, water nuts, biscuits etc. They are often associated with *Sitophilus* and *Tribolium*.

3. *Trogoderma granarium* Evert (Khapra beetle)

(Coleoptera : Dermestidae)

Temperature - 35 ± 2°C

Humidity - 70 ± 5%

Food - Wheat grains + Crushed wheat + Yeast in the ratio 40:6:1

Procedure

Take sterilized wide mouth glass jar and filled it half with sterilized food. Release the adult beetles, in equal sex ratio. Cover the jar with muslin cloth tightly held with rubber band.

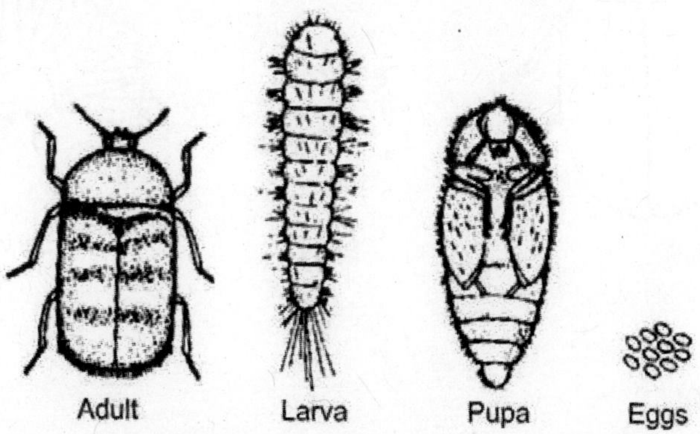

Fig. 6.8. Life history of *Trogoderma granarium*

Keep the culture in the laboratory at 35°C temperature and study the life cycle.

Its life history was studied by Atwal and Bains (1969-74) in India. This pest completes 4-5 generations from April to October and November to March.

The female lays 80-125 eggs. The larvae are active. move and feed freely. They are yellowish brown in colour and long hairs (bristles) are seen. Larvae are highly resistant to starvation. Pupal period lasts for 5-8 days. Adults are short lived for about 14 days. Life cycle is completed in 25 days under optimum conditions.

Note : Sexual dimorphism is well marked. The male is smaller than female. It attacks a wide variety of cereals, oil seeds, broken pulses etc.

4. *Lasioderma serricorne* (Fabricius) (Cigarette beetle)

(Coleoptera : Anobiidae)

Temperature - 30 ± 2°C

Humidity - 70 ± 5%

Food - Fennel seeds, Turmeric etc.

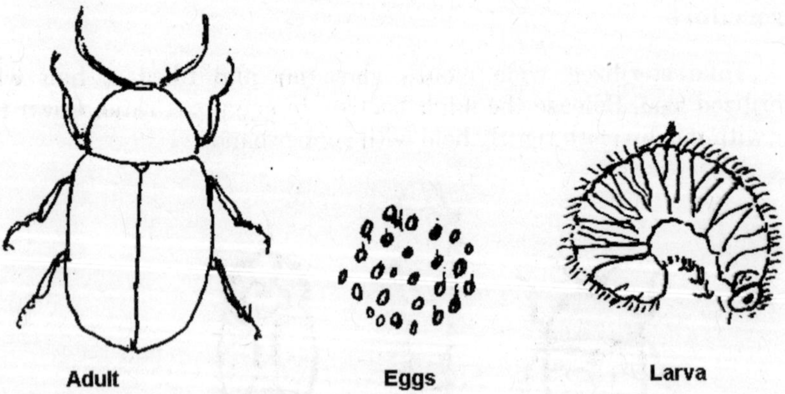

Adult Eggs Larva

Fig. 6.9.

Procedure

Take a wide mouth glass jar. Wash it thoroughly and sterilized it by keeping in oven at 60°C temperature for 24 hours. Food is also sterilize similarly and after cooling add 5 gm yeast. Release the adult beetles (male and female ratio 1:1) and cover the jar with muslin cloth held tightly with rubber band. Keep the culture in the laboratory at 30°C temperature and 70% RH. and study the life cycle.

Its detailed biology was worked out by Howe (1957). Female lays about 100 eggs. The eggs are white in colour. The incubation period is of about 6 days. The newly hatched larvae are very active and cut tiny holes in the seeds and after 4 days change into second instar which after 6 days metamorphose into 3rd instar (duration 7 days). After 8 days, the fourth instar larva pupates. After 6 days of pupal life, the adults emerge out. The life cycle completes in 36 days in optimum conditions.

Note : The males are smaller than females. It attacks a variety of articles including, tobacco, cereals, pulses, turmeric, copra, oil cake, condiments and spices, exporting mushrooms dried fish etc. It is very hard to rear it below 18°C temperature. The species can be confused with *Stegobium paniceum* which is somewhat similar in appearance and size. Larvae are negatively phototototactic. It can be reared on tobacco, turmeric and wheat flour also. (wheat flour + wheat bran + yeast in the ratio 50 : 50 : 5).

5. *Oryzaephilus surinamensis* (L.) (Saw toothed grain beetle)

(Coleoptera : Sylvanidae)

Temperature - 30-35°C

Humidity - 70-90%

Food - Rice grains or wheat flour.

Procedure

Take clean and sterile wide mouth culture jar. Filled it half with sterile rice grains that contain some crushed grains also. Release the adult beetles in equal sex ratio and cover the jar with muslin cloth hold tightly with rubber band. Keep the culture jar in the laboratory at 30°C temperature and 70% RH and study the life cycle.

Life cycle of *O. surinamensis* was studied by Howe (1956). Female deposit their eggs in crevices of the grain and dropped loosely in flour. The eggs are small and white. The larvae are elongated and make cocoons of gelatinous substance to which the food particles are adhered.

In optimum conditions the life cycle completes in 24 days, during the months June-July.

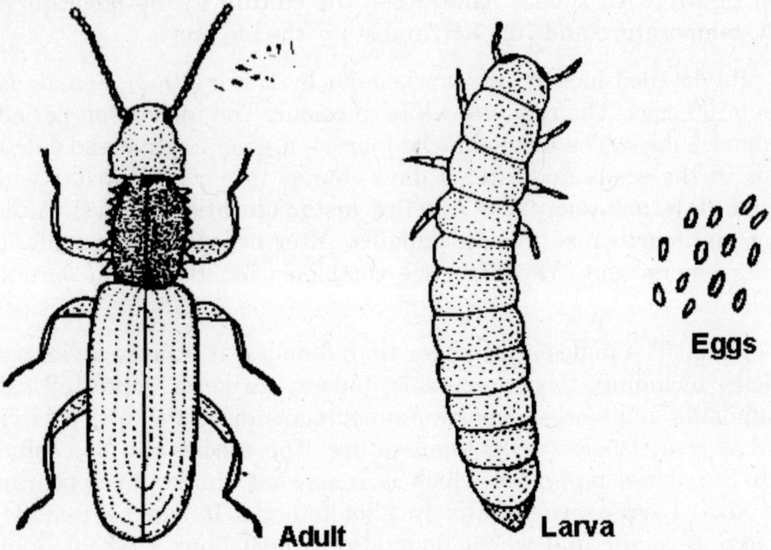

Fig. 6.10. *Oryzaephilus surinamensis* Linn. Adult, Larva, Eggs.

Note : They attack a wide variety of products like rice, flour, cornflex, poultry feed, dry fruits and nuts, candies, tobacco, snuff, yeast, dried meat etc. It is often associated with *Tribolium, Sitophilus* and *Oryzaephilus*. At 35°C temperature and 90% RH, the life cycle takes about 20-21 days.

6. *Corcyra cephalonica* (Staint) (Rice moth)

(Lepidoptera : Pyralidae)

Temperature - 30-32°C

Humidity - 60-65%

Food - Crushed sorghum.

Procedure

Wide mouth jar is cleaned and sterilized. Food is also sterilized. Adult moths are released in equal sex ratio and cover the container with muslin cloth held tightly with rubber band.

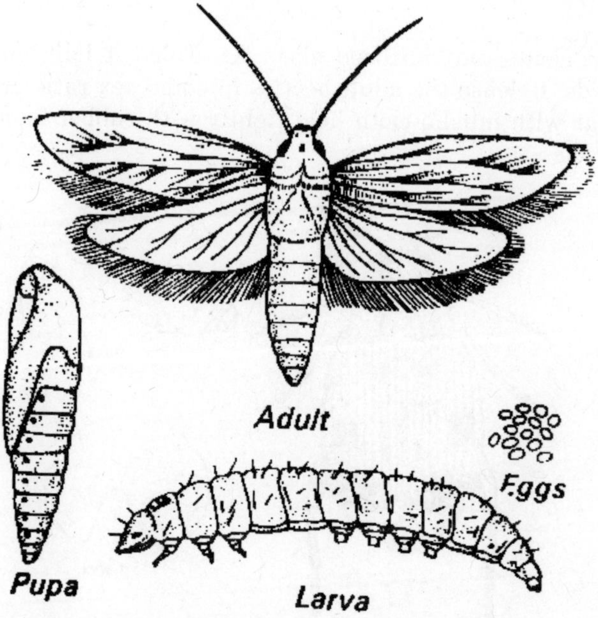

Fig. 6.11. Life history of *Corcyra cephalonica*

The eggs (100-150) are laid singly or in groups on the grains. The larvae produce silken tunnels under which they feed. Many tunnels can be seen in the culture jar housing a larva in each of them. Pupation takes place in these silken cocoons. After 9-14 days of pupal life, adult moth emerge out. Larval duration ranges from 21-41 days and incubation period of eggs is about 4-7 days. As many as 7 larval instars are found.

Note : It attacks grains like sorghum, maize, gram, ground- nuts and cotton seeds. The damage is done by larvae only. The male is smaller than female.

7. *Callosobruchus chinensis* (L.) Gram dhora

(Coleoptera : Bruchidae)

Temperature 30-32°C

Humidity - 70-90%

Food - Pulses and grams

Procedure

Take clean and sterilized glass jar. Filled it half with sterilized gram seeds. Release the adult beetles in equal sex ratio and cover the culture jar with muslin cloth held tightly with rubber band and study the life cycle.

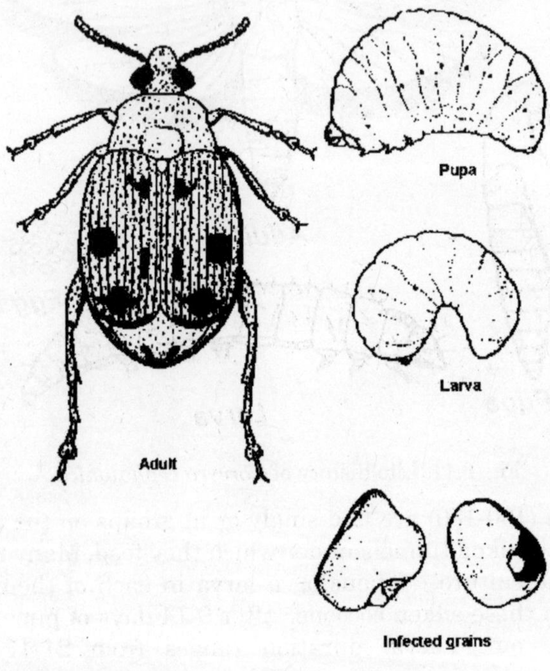

Fig. 6.12. Life stages of *Callosobruchus*

Its life cycle was studied by Howe and Curie (1964).

Female lays the eggs singly on the gram seeds (as many as 8 eggs could be glued to a single seed) *C. analis* lays only one egg per seed. The larvae on hatching bore into the seed, lie curled up and grow by eating the seed contents.

As many as 100 eggs are laid. The incubation period varies from 4-6 days. The larval duration varies from 14-21 days. It is active during March to November. Pupation takes about 7 days to complete.

Note: It attacks channa or chhola, masure, mung, arhar, pea, soyabean, urd etc. In males the antennae long and pectinate whereas it is sub-serrate and short in females. Elytral tubercle is absent in female but present at the bases of 3rd and 4th striate in male. It is associated with *C. maculatus* and *C. analis*.

8. *Poecilocerus pictus* (Aak grass hopper)

(Orthoptera : Pyrgomorphidae)

Temperature - 33 ± 2°C

Humidity - 60 ± 5%

Food - Fresh leaves of *Calotropis* plant.

Adults and nymphs of *P. pictus* can be collected from the calotropis plants in field during April to July and conditioned in the laboratory in large wooden cages measuring 24" x 24" x 25" in size.

Keep the adults of both sexes in cages measuring 11" x 7" x 9" size fitted with mesh window on the sides and glass door in front. The cage is fitted with separable wooden false floor with seven holes of 2" diameter under which the transparent glass tube of about 3" long filled with moist finely sieved sterilized sand are fixed for oviposition. The glass tubes containing eggs pods are removed and kept at 33 ± 2°C temperature and 60 ± 5% RH.

There are six nymphal instars in the life cycle. The longevity of the adult is about 25-30 days. There are two broods in year. 150-200 eggs are laid. The egg pod is laid deep under the soil. Incubation period of summer brood is about 30 days whereas the eggs laid in September-October are hatched in April. It takes hardly 25 days to become adult (see figure 3.8b).

9. *Periplaneta americana* L. (Cockroach)

(Dictyoptera : Blattidae)

Take a wooden cage or glass trough and put rolled brown paper or packing material (newspaper). Bread or dog biscuits are kept as food material. Release the adult cockroaches (males and females). Cover the trough with muslin cloth held tightly with rubber band. Clean the culture trough every fortnight so that the fungal infection could not take place.

The eggs are laid in a ootheca which is carried protruding at the end of the female abdomen for several days before being deposited.

The incubation period is about 20-21 days. As many as nine nymphal instars are found. The nymphal duration is of 25-40 days.

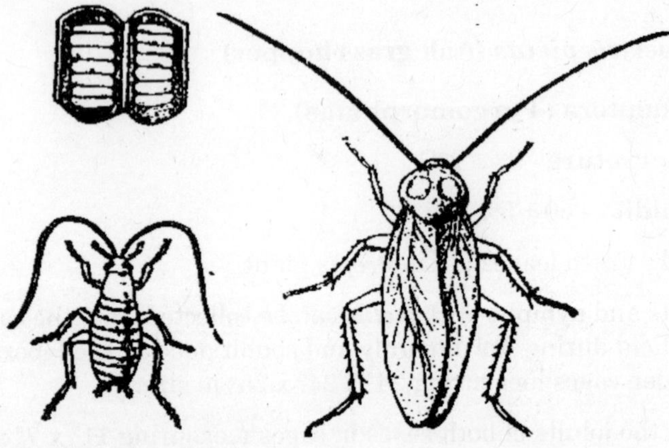

SECOND INSTAR NYMPH

Fig. 6.13. *Periplaneta americana* Linn.
Adult, Egg case with longitudinal rows of eggs, Second.

Note : The cockroaches are nocturnal therefore, it is advisable to keep the culture jar in dark or shaded corner.

10. *Musca domestica* Fabr. (Housefly)

(Diptera : Muscidae)

Temperature - 28 ±1°C.

Procedure

A wooden cage of 1' x 1/2' x 1-1/2' dimensions is required. The front side of this cage is made of sliding glass. On the other side a small sliding door is also there to keep the egg laying medium or liquid food etc. Small window covered with wire gauze is located on the back side of the cage. Release the adult flies in the cage.

They lay eggs in the medium which is kept in the petridish inside the rearing cage. The medium is made up of wheat bran + yeast soaked in water.

2-3 days after mating, the adult females lays eggs in the fermenting medium. It is then taken out and fresh medium for egg laying is kept. The eggs are hatched with in 24 hours and larval duration gets completed in 6-9 days. Mature larvae leaves the medium and pupates. The pupal stage lasts for about 5-6 days. Another culture can be set with these pupae (see figure 3.36a).

Note : *Musca nebulo* is a tropical species found in our houses whereas *Musca domestica* is a temperate species found in Himalayan region and temperate regions of the world.

11. *Sarcophaga* sp. (Flesh flies)

(Diptera : Sarcophagidae)

Procedure

Take a rearing cage of 2' x 2' x 3' dimensions made up of fine cage, one with meat piece and another with sterile moist soil for pupation. Collect some flesh flies from near the slaughter houses and release in the cage. They lay eggs on the meat. Eggs hatch out into maggots which after several instars crawl out and pupate in the damp soil. The pupae can be collected to set the fresh culture.

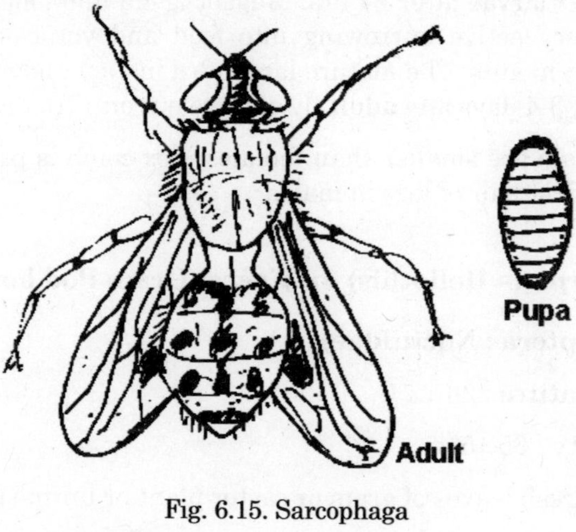

Fig. 6.15. Sarcophaga

12. *Drosophila melanogaster* (Fruit fly)

(Diptera : Drosophilidae)

Temperature - 20-25°C

Artificial diet	- Agar agar -10gm	- (A)
Non crystal sugar	- 65 gm.	- (B)
Corn flour	- 75 gm.	
Yeast powder	- 10 gm	

Dissolve the agar-agar into 300 ml. of water and warm till it dissolves completely (A)

Mix the ingredients (B) into 450 ml. of water. Stir and Mix both (A) and (B). Add 2 ml. propionic acid as antifungal agent. Pour it in culture bottles.

Procedure

Take the milk bottles (200-300 ml. capacity) and filled them 1/4 with the artificial diet. Release the adult flies (males and females) in the culture bottle and plug with cotton. They mate and lay eggs on the surface of the medium.

The egg is whitish with two anchors or filament. The eggs hatches into 1st instar larvae after 24 hrs. Maggot is 4.5 mm long creamish in colour and very active burrowing into food and voracious feeder. It undergoes two moults. The mature larva (3rd instar) undergo pupation. Pupa lasts for 3-4 days and adult fly emerge out on 11th day.

Note : Male are smaller than females. Sex comb is present on the tarsus of the first pair of legs in males.

13. *Helicoverpa* (= Heliothis) *armigera* (Gram Pod borer)

(Lepidoptera : Noctuidae)

Temperature : 28 ±2°C

Humidity : 65 ±5%

Food : Fresh leaves of gram or castor plant or unripe tomato.

For artificial diet see appendix I.

Procedure

Collect the larvae of *Helicoverpa* from the gram field and rear them in the laboratory. It is advisable to keep the mature larvae in separate plastic vials as they exhibit cannibalism. Make some holes in the lid of the vial by hot needle for aeration. Clean the vial daily. Four larval instars are found. The pupa is dark brown which after 9-14 days emerges into adult. The adult are provided with 5% sugar solution (cotton soaked in it) in a petridish. They are kept in glass chimneys for eggs laying where the paper strips hung to settle and lay eggs. Eggs are spherical yellow in colour.

Note : Male moth is smaller than female and has dark spots on wings.

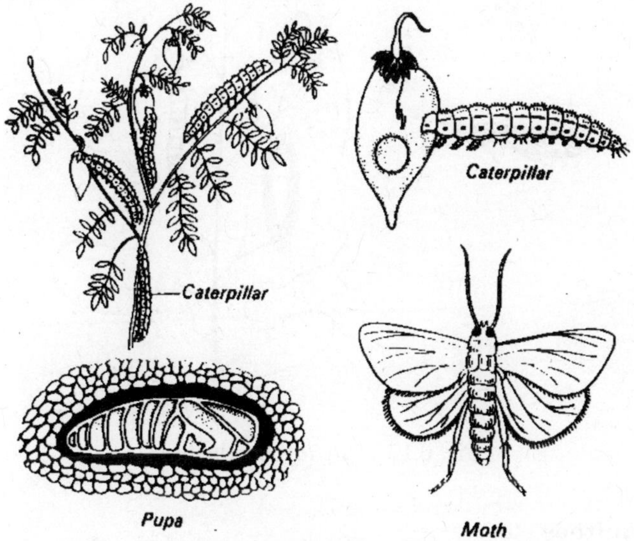

Fig. 6.16. Life history of *Helicoverpa armigera*

14. *Plutella xylostella* (Diamond back moth)

(Lepidotera : Plutellidae)

Temperature : 30-32°C

Humidity : 70-75%

Food- Fresh leaves of cabbage

Collect the larvae from the fields and rear them in laboratory. The pupation takes place in silken cocoon. It lasts for 9-14 days. Adults are small moth, the female is bigger in size than male. They are kept in lantern globe where they mate and lay eggs on the leaf surface of the cabbage plant. Lantern globe is placed inverted on the small plants.

Eggs are hatched in 2-3 days. First instar larva is very small and change into second instar after 2 days. Five larval instars are found. They cause extensive damage to the leaves and cut holes in them make galleries. In ideal conditions the life cycle is completed in 20-22 days whereas in winter it takes 25-28 days.

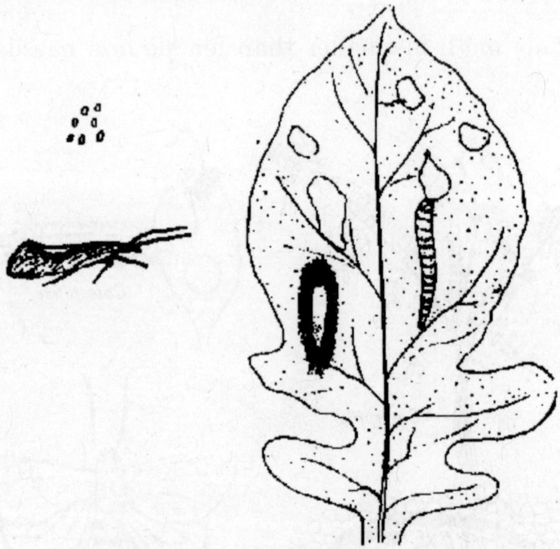

Fig. 6.17. *Plutella* - Life history

15. Mosquitoes

(Diptera : Culicidae)

Temperature - 28 ± 2°C

Procedure

Collect the larvae from the water with the help of strainer and transfer them in separate enamel trays filled with water. A pinch of yeast powder can be spread as food for the larvae. They change into pupal stage within 10-15 days. Pupa after 3-4 days transform into adult. The pupae are separated from the enamel tray and are kept in other trays kept inside the cages of muslin cloth on a wooden or iron frame so that the adults could not be escaped. Adults requires blood meal. The eggs are hatched in 2-3 days. The complete life cycle takes 20-25 days. The rearing of mosquitoes can be done at room temperature except winters.

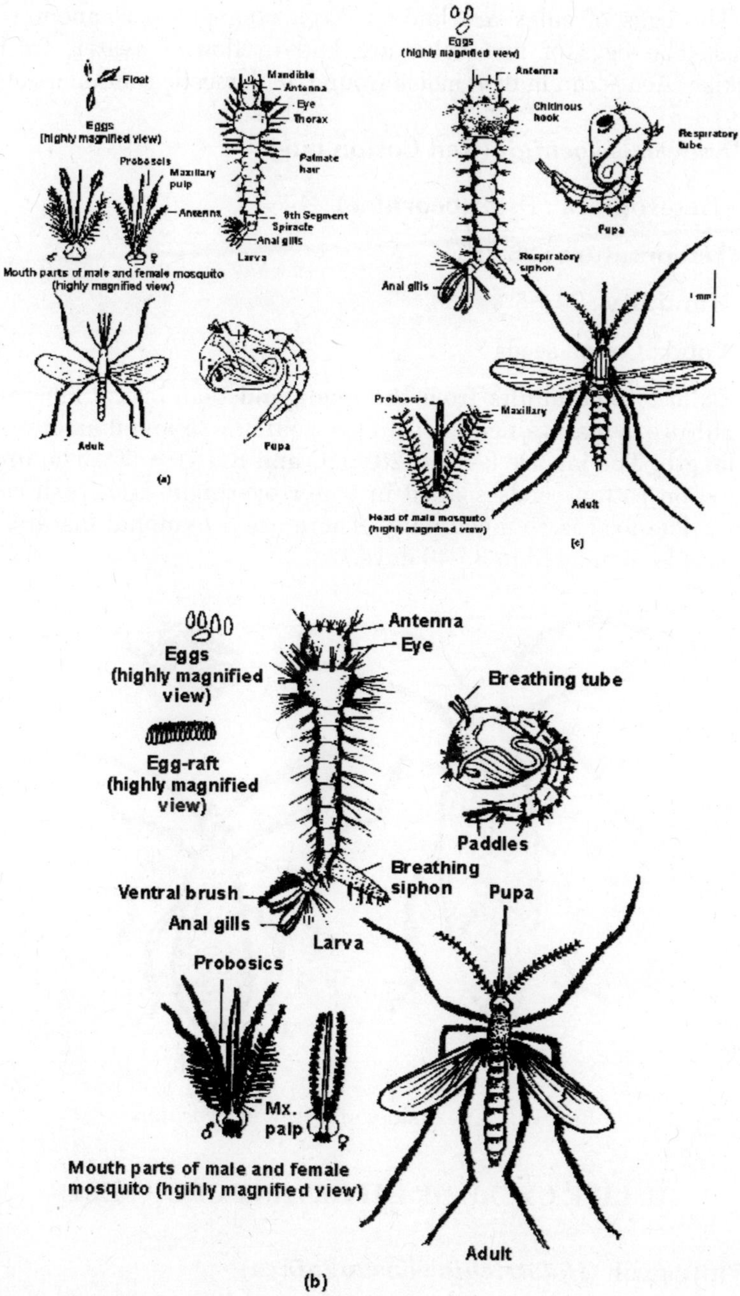

Fig. 6.18 (a, b & c). Life cycle of *Anopheles, Aedes* and *Culex mosquitoes*

The eggs of culex are laid in dirty, stagnent permanent water bodies. The eggs of anopheles are laid in flowing water, fresh or brackish. Aedes can laid on moist ground, in domestic containers etc.

16. *Dysdercus koenigii* (Red Cotton bug)

(Heteroptera : Pyrrhocoridae)

Temperature : 28 ±1°C

Humidity : 70 ± 5%

Food : Cotton seeds

Collect some adults from the fields and rear in the laboratory. The culture is maintained in clean glass jars in 25 mm diameter x 20 cm. length. The jar are kept at 28 ±1°C and RH 70 ± 5%. The insects are fed on cotton seeds soaked in water overnight and fresh cotton leaves. The food is changed daily. There are 5 nymphal instars. The life cycle is completed in 35-40 days.

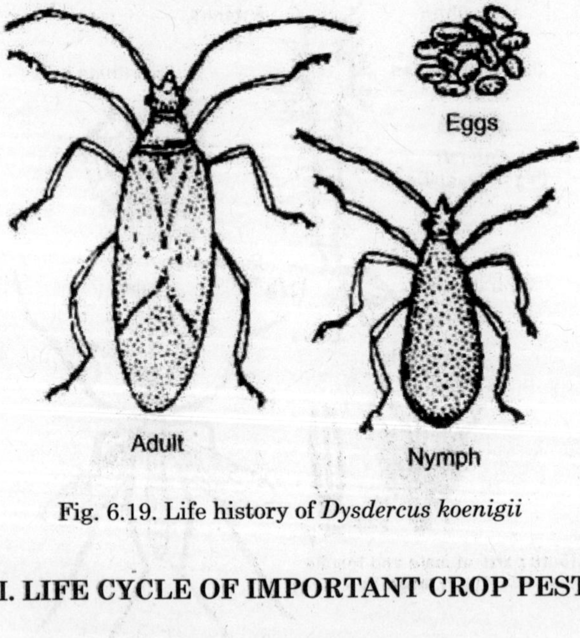

Fig. 6.19. Life history of *Dysdercus koenigii*

III. LIFE CYCLE OF IMPORTANT CROP PESTS

1. White grub (*Holotrichia consanguinea*)

(Coleoptera : Melolonthidae)

Identification : Copper coloured, large size, nocturnal beetle.

Host plants : Ber, Neem, Gular, Sainjana, Jamun. It is a polyphagous pest.

Crops attacked : Groundnuts, Millets, Maize.

Damage : Both adult and grubs cause damage. Adults defoliate the plants during night whereas grubs feed on root and rootlets of variety of crops.

Life cycle : Female starts laying the eggs after 2-3 days of mating. The eggs are hatched in 7-13 days. The first instar larva remains 13-19 days. Three instars are found. The pupa lasts from 11-18 days. The life cycle is completed in 82-113 days. It is active from June to November.

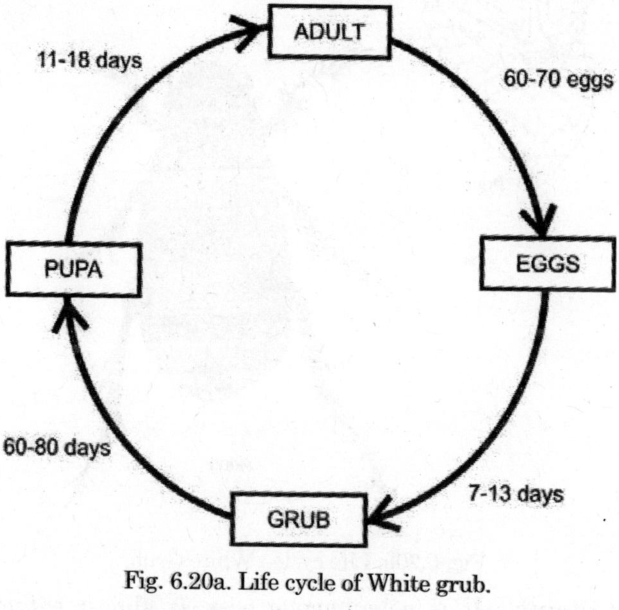

Fig. 6.20a. Life cycle of White grub.

2. Desert Locust (*Schistocerca gregaria*)

(Orthoptera : Acrididae)

Identification: It occurs in two phases. The adults of *gregarious phase* are yellow in colour when sexually mature (pink when immature) whereas the *solitary phase* is grey or greenish. Pronotum is saddle shaped, gregarious phase fly during the day in swarms.

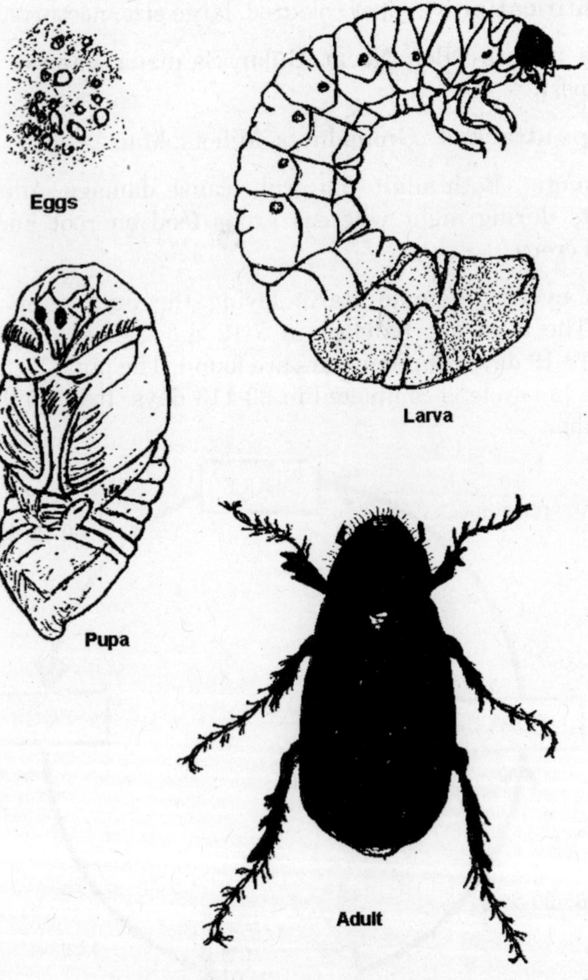

Fig. 6.20b. Life cycle - White Grub.

Host-plants : It is polyphagous pest. It almost eat everything except Neem, Dhatura etc.

Damage: Both nymph and adults cause damage. It cause extensive damage when in swarm.

Life Cycle : Eggs hatch in 12-15 days in summers (3-4 weeks in winters). 5 nymphal instars are found. They take 3-4 weeks to develop into adult in summers whereas 6-8 weeks in winters (Fig. 6.21).

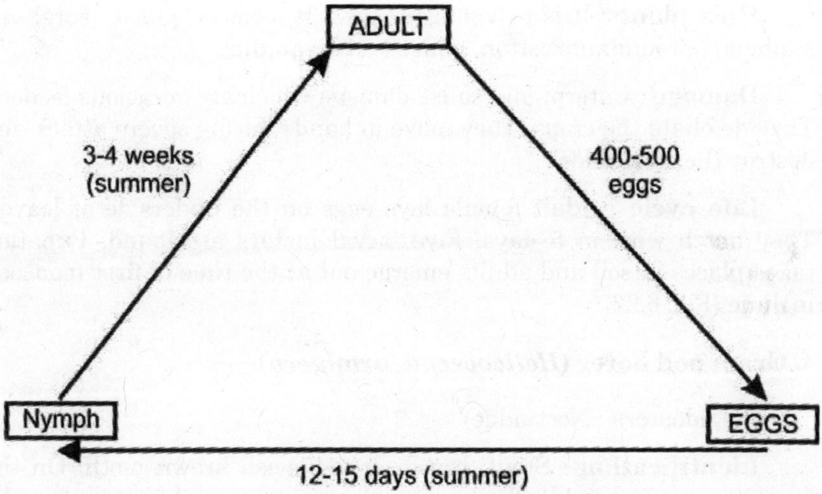

Fig. 6.21. Life cycle of Locust.

3. Red Hairy caterpillar (*Amsacta moorei*)

(Lepidoptera : Arctiidae)

Identification: White wings bear black spots. The outer margins of the forewings, part of the thorax and abdomen is scarlet red in colour.

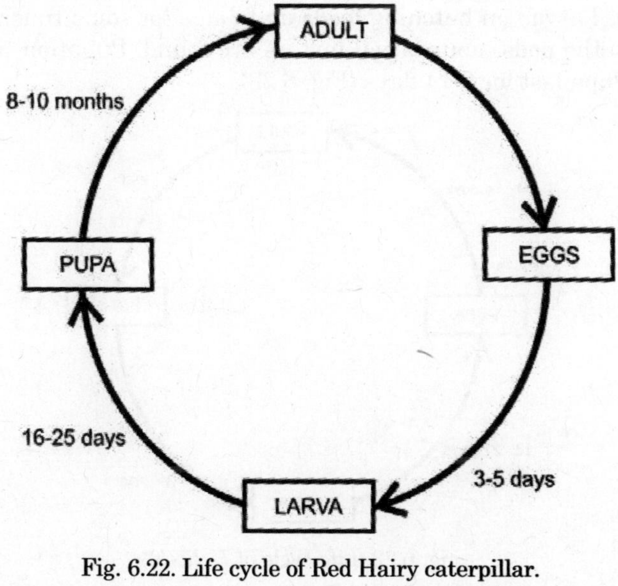

Fig. 6.22. Life cycle of Red Hairy caterpillar.

Host plants: It is polyphagous pest. It feeds on pulses, sorghum, sunhemp, groundnuts cotton, maize and sesamum.

Damage : Caterpillars cause damage which are voracious feeders. They defoliate the crops. They move in bands during severe attack and destroy the crop fields.

Life cycle : Adult female lays eggs on the underside of leaves. They hatch with in 5 days. Five larval instars are found. Pupation takes places in soil and adults emerge out at the time of first monsoon in June (Fig. 6.22).

4. Gram pod borer (*Helicoverpa armigera*)

(Lepidoptera : Noctuidae)

Identification : Stout, brown or yellowish brown moth. On the forewings are located dark specks and a conspicuous black spot in the middle of the wings. Hind wings are dull coloured with black border.

Host plants : Larvae are polyphagous yet they prefer gram and peas.

Damage : Larvae feeds on the green parts defoliating young crops. Extensive damage is caused to pods where they cut holes in them and consume the seeds.

Life cycle : Adult female lays eggs in the foliage. They hatch in 3-5 days. Larvae on hatching feeds on foliage for some times and later bore into the pods. Four larval instars are found. Pupation takes place in soil. Pupa last for 9-14 days (Fig. 6.23).

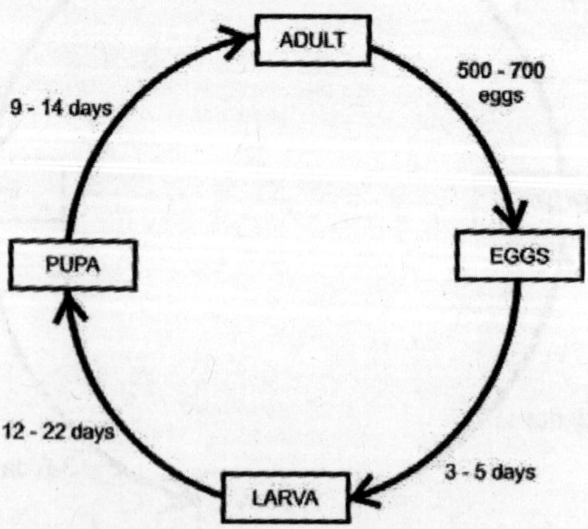

Fig. 6.23. Life cycle of *Heliothis*

5. Oriental fruit fly (*Dacus dorsalis*)

(Diptera : Tephritidae)

Identification : Stout fly, brown in colour, wings transparent and legs yellow.

Host plants : A number of fruits are attacked viz. mango, guava, plum, cherry, cheeku, ber, citrus, apricot, brinjal, peach etc.

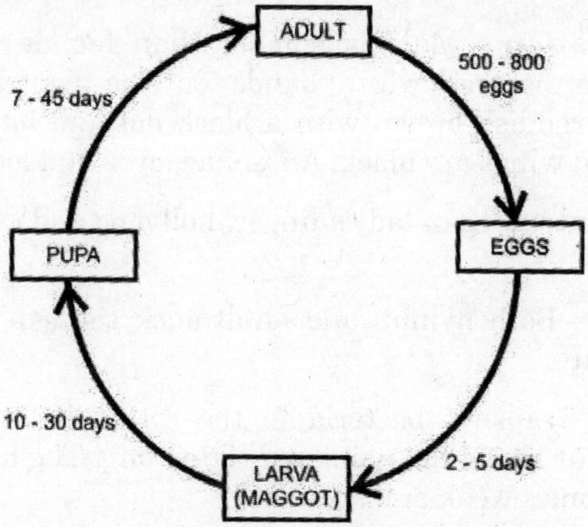

Fig. 6.24a. Life cycle of Oriental fruitfly.

Damage : Extensive damage is caused by maggots. Adult feeds on exudation of ripe fruits. Female insect punctures for egg laying that leads to secondary infections by microorganisms. Maggots destroy pulp and its excreta inside the fruits cause rotting.

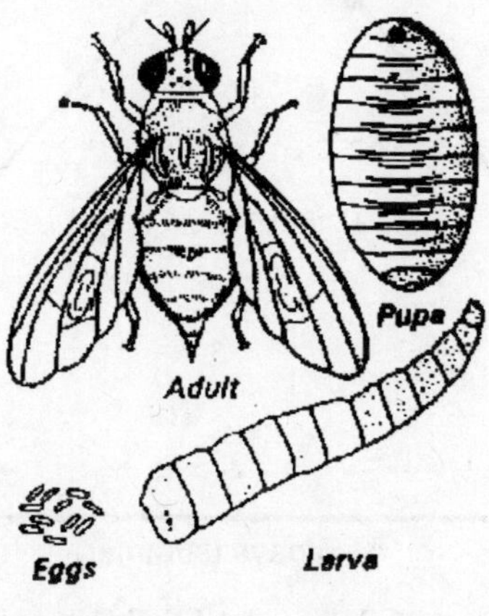

Fig. 6.24b. Life history of *Dacus dorsalis*.

Life cycle : The eggs are laid in clusters of 2-15. They hatch within 2-5 days. Three larval instars are found that took 10-30 days. Pupation takes, place in soil. Adult flies emerge in April (Fig. 6.24a & b).

6. Red cotton bug (*Dysdercus koenigii*)

(Heteroptera : Pyrrhocoridae)

Identification : Male is smaller than female. Body is red in colour with transverse white bands on the ventral surface. The forewings are reddish brown with a black dot and half of the wing is black. The hind wings are black. Antennae, eyes and legs are black.

Host plants: cotton, lady's finger, hollyhoc and other Malvaceous plants.

Damage : Both nymph and adult suck the sap (juice) from the leaves and fruits.

They can transmit bacteria in the cotton ball that causes red colour of the lint hence the name *red cotton bug*. The leaves are curled, the plants becomes weak and stunted.

Life cycle : The eggs are laid in clusters in the crack of the soil or dry leaves. Hatching takes place with in 4-5 days. Five nymphal instars are found that lasts for 30-35 days. A single female can lay as many as 400 eggs (Fig. 6.25).

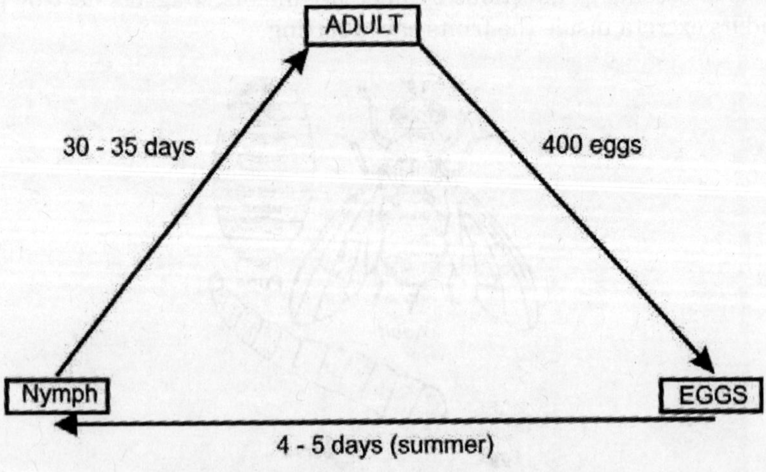

Fig. 6.25. Life cycle of Red Cotton bug.

7. Sugarcane leaf hopper (*Pyrilla purpusilla*)

(Hemiptera : Fulgoridae)

Identification : Head beak like pointed size about 20 mm. long and straw coloured. Female bear a pair of anal tuft.

Host plants : Sugarcane, wheat, maize, oat, millets and barley.

Damage : Nymph and adult both cause damage by sucking sap from the foliage. Honey dew secreted by these attracts fungi that inter fere with photosynthesis. Leaves turns black due to the accumulation of sooty mould. Leaves fall down.

Life cycle : Female lays 600-700 eggs that hatch after a week or so. Five nymphal instars are found which after 40-60 days, metamorphose into adults (Fig. 6.26).

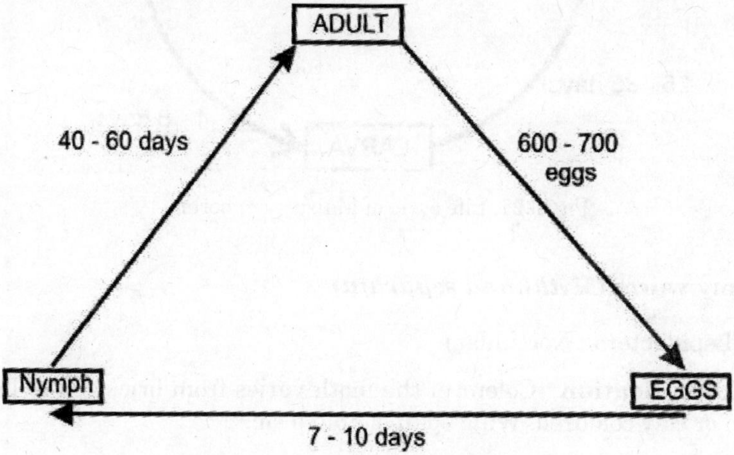

Fig. 6.26. Life cycle of Pyrilla.

8. Maize stem borer (*Chilo partellus*)

(Lepidoptera : Pyralidae)

Identification : Fore wings of the moth are light brown in colour with numerous shining brown spots on the margin. Hindwings are white and thin.

Host plants : maize, jowar, bajra.

Damage : Caterpillar damage the stem by eating it from inside.

Life cycle : The female lays the eggs in cluster of 50- 150 during April-May in rows under the leaves. Hatching takes place within 4-8 days. Five larval instars are found that takes 25-35 days to go under pupation. Pupa lasts for 9-14 days. Adults emerge out and after mating lays 250-300 eggs. It is active during April to October (Fig. 6.27).

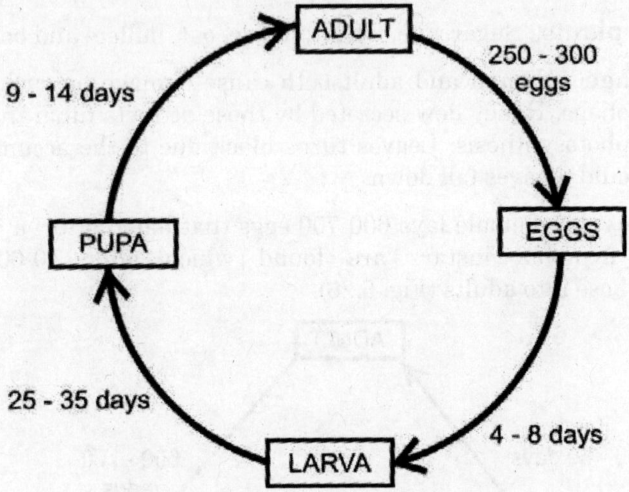

Fig. 6.27. Life cycle of Maize stem borer.

9. Army worm (*Mythimna separata*)

(Lepidoptera : Noctuidae)

Identification : Colour of the moth varies from brick red to light brown or clay coloured. Wing span is upto 5 cm.

Host Plants : wheat, maize, sugarcane, paddy and some other plants of the family Graminae.

Damage : Caterpillars cause extensive damage. In case of severe attack, the whole leaves are eaten up. The caterpillars suspend themselves by silken threads and blown by the wind to infect new plant.

Life cycle : Female moth lays eggs in rows or clusters on dry leaves. Hatching takes place from 3-12 days. There are six larval instars that pupates after 12-20 days. Pupation takes place in soil at certain depth (0.5-5 cm). As many as 100 eggs are laid by the female (Fig. 6.28).

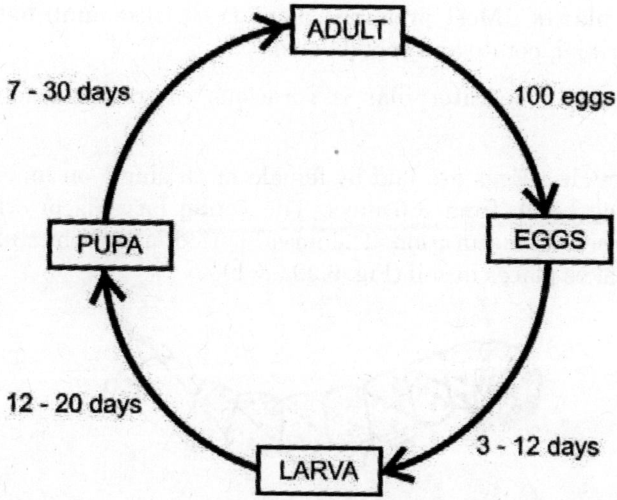

Fig. 6.28. Life cycle of *Mythimna*

10. Til Hawk moth or Death's head moth (*Acherontia styx*)

(Lepidoptera : Sphingidae)

Identification : Adult moth is large, robust and reddish brown in colour. The thorax bears a prominent death's head mark. The forewings have dark brown and grey pattern with dark wavy markings.

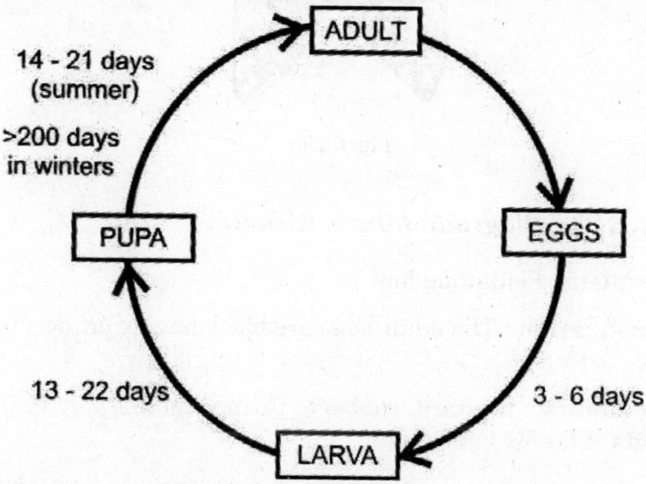

Fig. 6.29a. Life cycle of *Acherontia*

Host plants : Most preferred plant is til (sesamum) but is also found on brinjal, potato and sem (beans).

Damage : The caterpillar is voracious eater that defoliate the plant.

Life cycle : Eggs are laid by female moth singly on underside of leaves. They hatch from 3-6 days. The young larva is brown which becomes green on maturation. The larval period lasts from 13-22 days. Pupation takes places in soil (Fig. 6.29a & b).

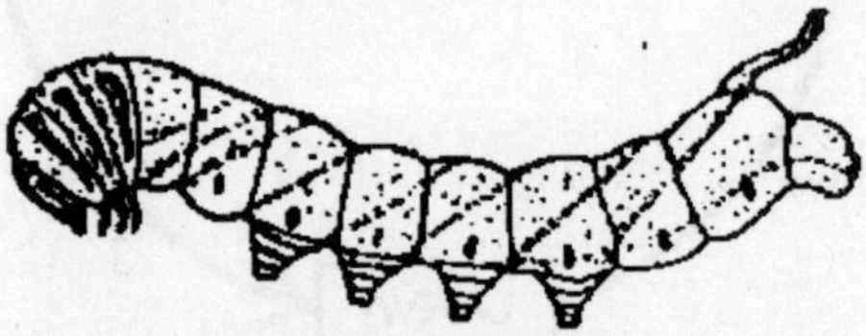

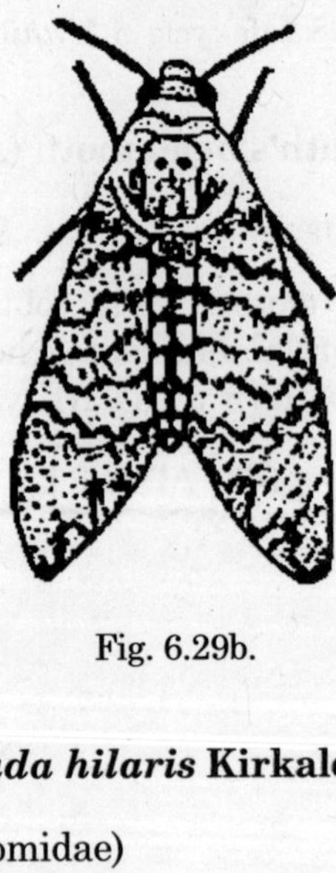

Fig. 6.29b.

11. Painted bug (*Bagrada hilaris* Kirkaldy)

(Hemiptera : Pentatomidae)

Identification : The adult bugs are black having orange or brown markings.

Host plants : mustard, cabbage, turnip, cauliflower, radish and other plants of family Cruciferae.

Damage : Adult and nymphs both suck the sap from the plant. The plants look stunted, dries up.

Life cycle : The female lays eggs singly on leaves, stem and leaf stalks. The hatching takes place from 3-15 days. Five nymphal instars are found.

The nymphal duration lasts for 15-21 days. In suitable weather conditions as many as 6-7 generations can be completed (Fig. 6.30).

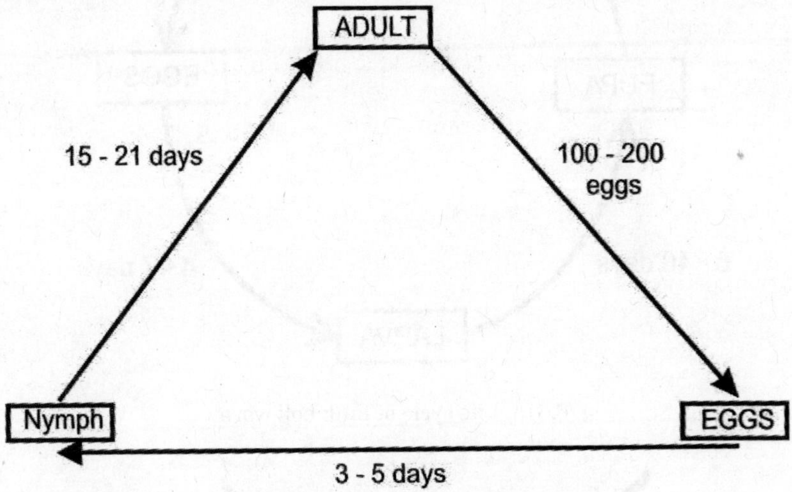

Fig. 6.30. Life cycle of Painted Bug.

12. Pink boll worm (*Pectinophora gossipiella*)

(Lepidoptera : Gelechiidae)

Identification : Small size moths with hind wings fringed and forewings deep brown with black spots.

Host plants : Cotton.

Damage : Larvae feeds inside the boll that causes premature shedding of the boll. Due to larval silk the bolls fails to open. Seeds are destroyed. Quality and quantity of lint declines.

Life cycle : Eggs are laid singly on tender leaves. On hatching they are white but soon turn pink and bore into the bolls. Three larval instars are found. Pupation takes place in fallen leaves or debris (Fig. 6.31a & b).

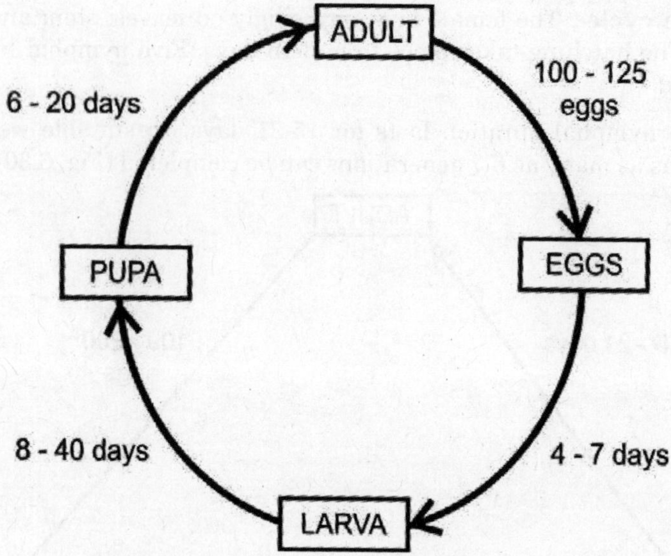

Fig. 6.31a. Life cycle of Pink boll worm.

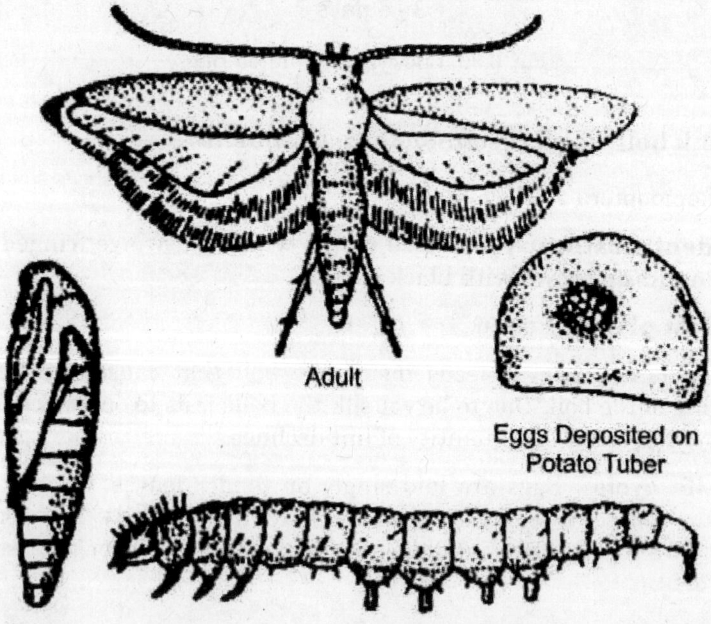

Fig. 6.31b. Life history of *Pectinophora gossipiella*.

13. Red Pumpkin Beetle *Raphidopalpa* (=*Aulacophora*) *foveicollis*

(Coleoptera : Chrysomelidae)

Identification: Adult 6-7 mm in size, orange red in colour, underside black.

Host plants: All cucurbitaceous fruits and vegetables viz. Pumpkin, Sponge gourd, Bottle gourd, Cucumber, Musk melon, etc.

Damage: The adult beetles feeds on leaves, buds and flowers and make holes. The grubs cause serious damage. They bore into roots, underground stems and sometimes in fruits. The plants are defoliated and flowers are shed.

Life cycle: 250-300 eggs orange/yellow in colour, are laid in moist soil near the base in the beginning of March upto October. The grub remains under the soil and feeds on roots, creeping stems and fruits lying contact with soil. Four instars are observed and takes around 122-22 days. Pupation takes place under the soil deeper into a water proof earthen chamber. The pupa after 7-14 days transform into adult.

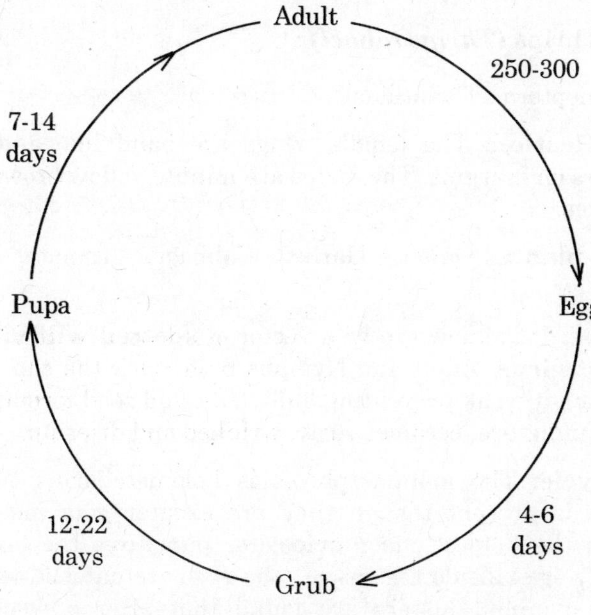

Fig. 6.31c. Life cycle of Red pumpkin beetle

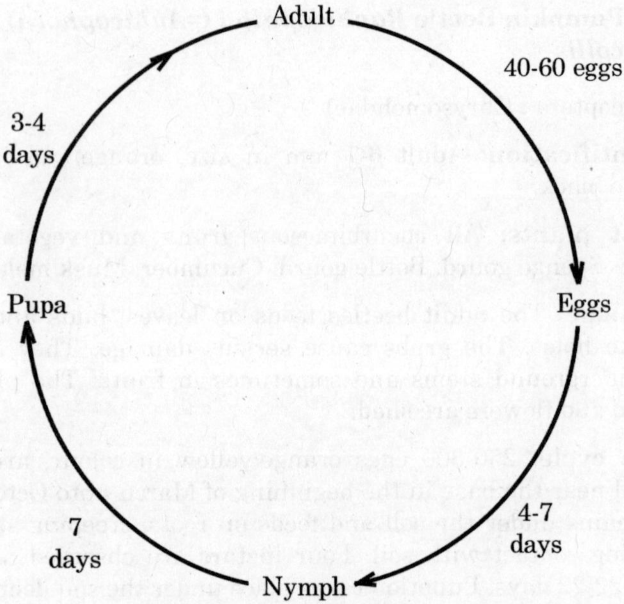

Fig. 6.31d. Life cycle of Thrips

14. Onion Thrips (*Thrips tabaci*)

(Thysanoptora : Thriptidae)

Identification: The female wings are band like and fringed bearing hairs on margins. The males are minute yellow brown insects without wings.

Host plant: Onion, Garlic, Cabbage, Tomato, Turnip, Cauliflower etc.

Damage: It is known to be a vector of **spotted wilt virus,** and **yellow spot virus**. Adult and Nymphs both, suck the sap from the plants making it weak preventing bulb, fruit and seed formation. The leaves after puncture, becomes curly, wrinked and dries up.

Life cycle: The metamorphosis is holometabolous types and pupal stage is present though they are exopterygote insects. The female with the help of sharp ovipositor punctures the leaves and stem and lay eggs inside the tissue. The eggs are hatched within 4-7 days. Four nymphal instars are found that after a week period transform into pupa. Full grown nymphs drops down to the soil to pupate around 5 cm. deep.

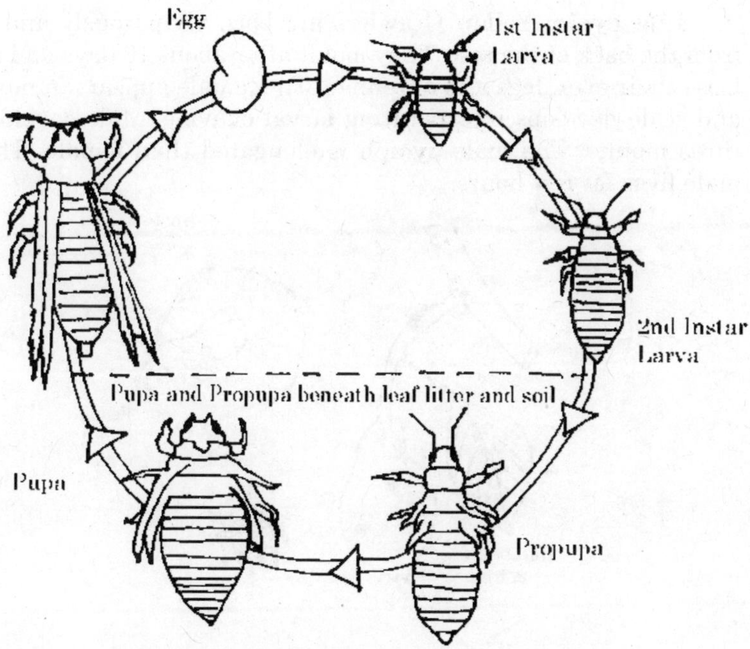

Fig. 6.31e. Thrips: Generalized life cycle

15. San Jose Scale* (*Quadraspidiotus perniciosus*)

(Hemiptera : Coccoidea : Diaspididae)

Identification: The adult male is orange in colour and has wings. The body of the adult female is yellow and is covered with 2 mm rounded dark gray scale.

Host plant: Major pest of fruit trees, Peer, Apple and several others and deciduous trees.

Damage : It is found on trunk, leaves, twigs, branches and fruits.

---

* San Jose Scale derives its name after in 1881 San Jose State, California (USA) where Comstoc discovered it is 1881. The species was originated in Siberia and north east China and spreads to all continents.

Life cycle: Yellow Crawlers are born viviparously and emerge from the back of the test. They moult after about 10 days and begin to loose their eyes, legs and antennae. The female appears in next moult and scale develops incorporating larval exuviae. Male develops after three moults. The male nymph is elongated than female. The adult male lives for few hours.

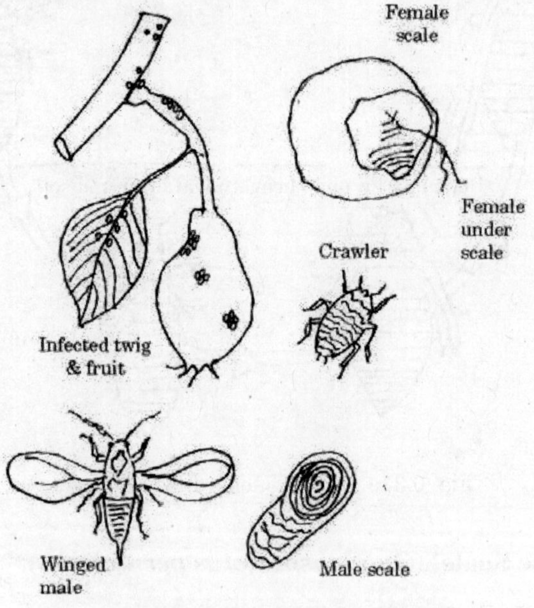

Fig. 6.31f. Life cycle of San Jose Scale

16. Cottony Cushion scale* (*Icerya purchasi*)

(Homoptera : Margarodidae)

Identification: The conspicuous feature is white cottony egg sac attached with body. The female is orange, yellow or brown. 600-800 eggs are found in egg sac.

---

* Cottony Cushion scale was discovered by Maskell in 1878 from a specimen sent to him by Dr. Purchas from Auckland, New Zealand One other species was known at that time in genus Icerya hence it was named *Icerya purchasi* for Dr. Purchas. It was native of Australia and spread to California. In India it came from San Jose State, California, USA in citrus and apples.

Host plants: Primarily on citrus and few ornamental plants.

Life cycle: Hermaphrodite, female is capable of self fertilization. Females are produced from fertilized eggs and males from unferilized. The crawlers that emerge from white sac of the female, starts life cycle. These crawlers settles on twig and leaves and moults twice. Three nymphal instars are found. The 3rd instar moults into adult female. The femalcs can reproduce without mating. Life cycle takes about three months to complete.

(1) Damage: They produce honey dew that develops moulds. The twigs dies and yield is reduced. They suc phloem tissue and reduce the vitality of host. *Rodolia cardinalis* can effectively control it.

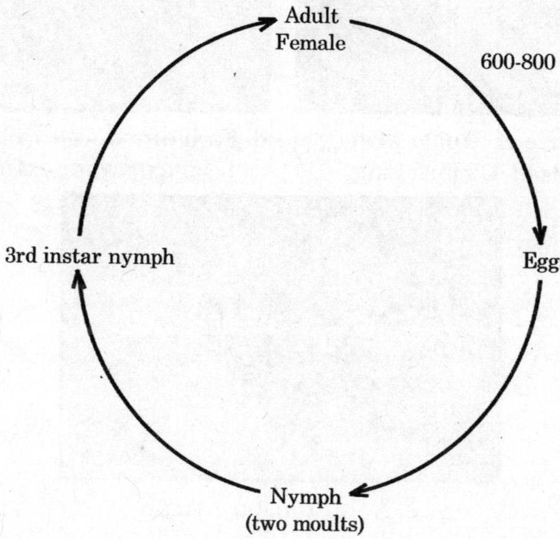

Fig. 6.31g. Life cycle of Cottony Cushion Scale.

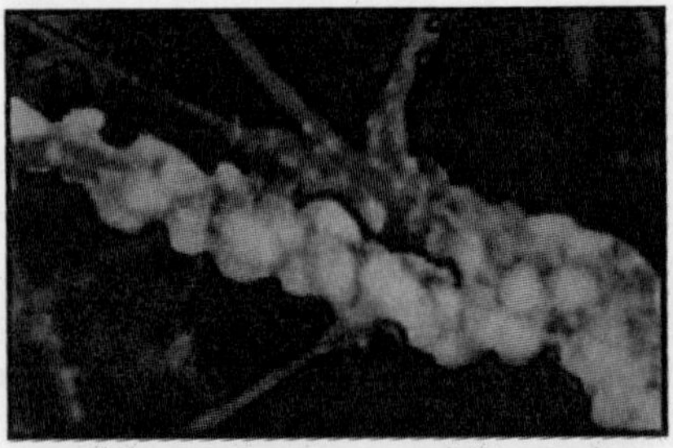

a. Colony of cottony cushion scale

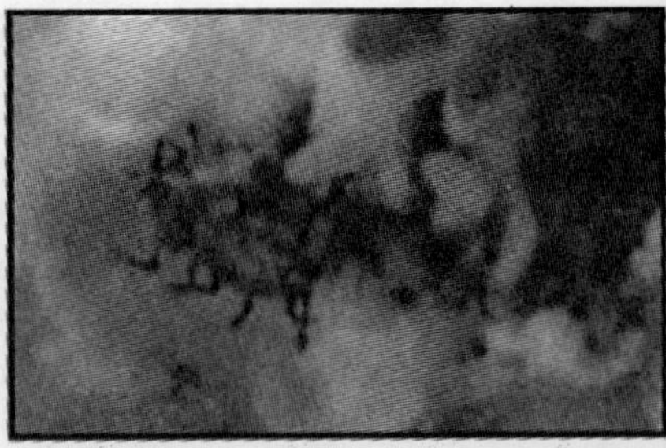

b. Cottony cushion scale crawlers

c. The actual size of female with egg sac can be about 1/2 inch long

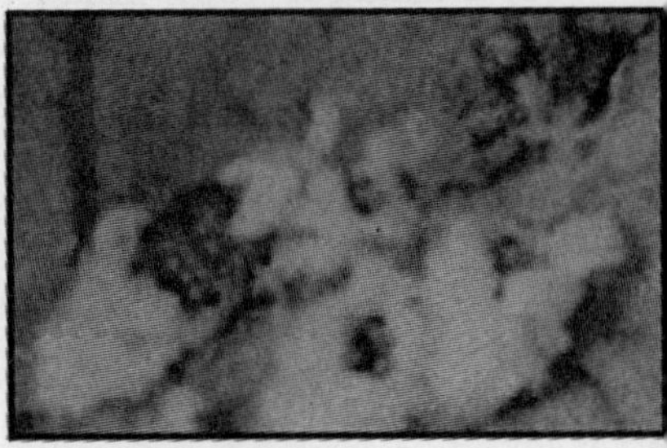

d. Each time a scale molts, it leaves behind its white, cottony molting skin

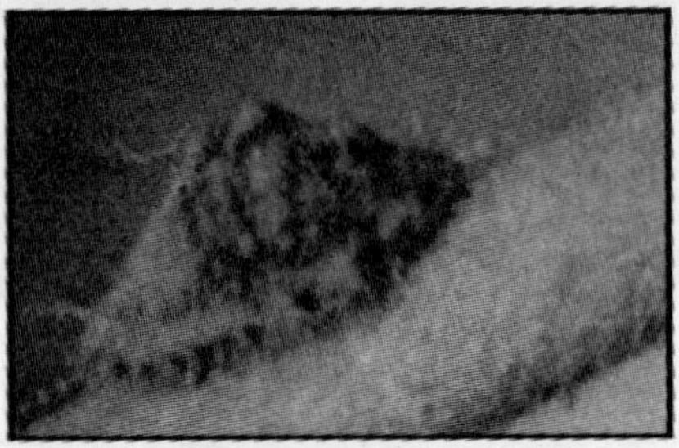

e. A third-instar nymph

Fig. 6.31h. Stages in life cycle of Cottony Cushion Scale.

BIOASSAY AND TESTING OF INSECTICIDES

Bioassay may be defined as a test in which the quantity or strength of material is determined by the reaction of a living organism to it. The purpose of bioassay is to answer the questions whether the compound is toxic? How much toxic? Does it vary in toxicity? Is it biodegradable? etc. If no, how much toxic?

For Bioassay study or testing we prefer to conduct acute, short term test. The current trend is to use the symboml LC_{50} or LD_{50} Lethal

Concentration or Lethal Dose which cause 50% mortality and 50% survival for a specific time. Therefore we use LC_{50}-24 hrs or LC_{50} - 48 hrs or LC_{50}-96 hrs. or LD_{50}-24 hrs or LD_{50}-48 hrs or LD_{50} -96 hrs. In aquatic medium we use the term LC_{50}.

Toxicologists are generally concerned with quantal and graded response. The quantal response is called "all or none" response i.e. it either happens or it does not happen. Graded response on the other hand, is continuous and is determined quantitatively. Toxicologists generally prefer quantal response in which LC_{50}, LD_{50}, EC_{50}, etc. are used.

Most assays design are based on a set of three to seven dose levels. Which are chosen on the basis of previous study or preliminary pilot studies, so that the LD_{50} will be as near the centre of the dosage levels as is possible. The dose-mortality curve can be approximated by a straight line, atleast in the neighbourhood of the LD_{50} when the mortality at given doses is plotted against the logarithm of the dose. For this reason, the doses are chosen in geometric progression e.g. 2, 4, 8, 16 units so that the logarithm of the doses increase by equal increments.

Given below is an example in which 60 insect were treated at each of the 4 doses.

S.No.	Dose (%)	No. of dead insects	No. of survived insects	Sg	Ds	Sg + Ds	$\frac{D_s}{S_g + D_s}$
1	1	10	50	121	10	131	$\frac{10}{131} = 0.08$
2	2	20	40	71	30	101	$\frac{30}{101} = 0.29$ b
3	4	31	29	31	61	92	$\frac{61}{92} = 0.66$
4	8	58	02	02	119	121	$\frac{119}{121} = 0.98$

Sg = Survival at this dose and greater doses; D_s = Death at this dose and the smaller doses; Mortality below 50% = Log b; The dose at which the mortality 50% i.e. 0.29 is 2%

Formula

$$\text{Log } LD_{50} = \text{Log b} + \frac{0.50 - \text{Mortality below 50\%}}{\text{Mortality above 50\%} - \text{Mortality below 50\%}} \times \text{Log}_2$$

Log 2 = Multiplication factor, the dose at which the mortality just below the required one, takes places

$$\text{Log } LD_{50} = 0.301 + \frac{0.50 - 0.29}{0.66 - 0.29} \times \log 2$$

Log b = Mortality below 50%.

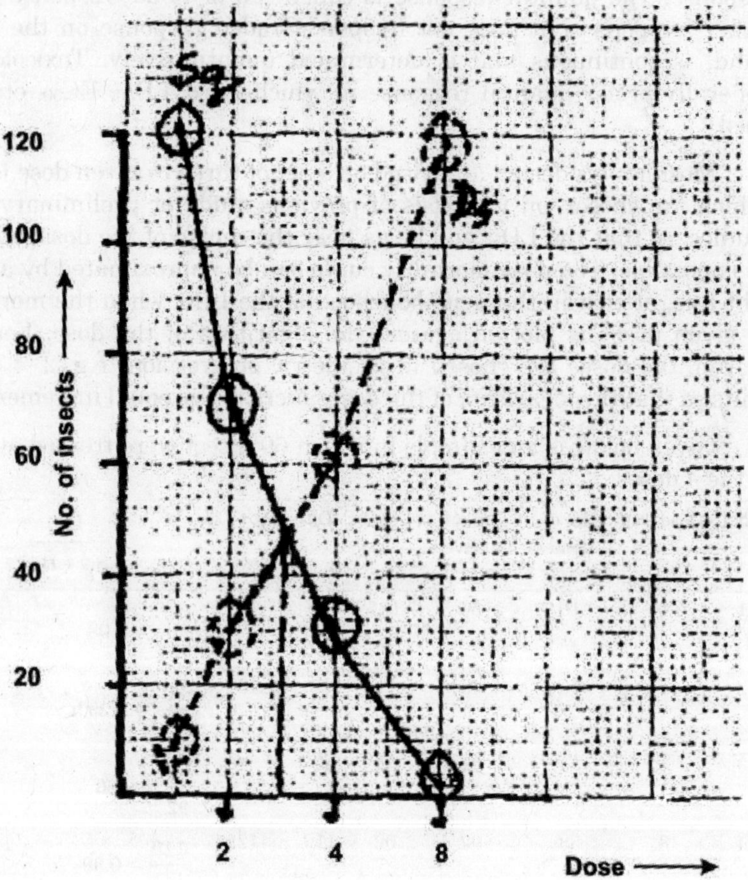

Fig. 6.32a.

$$= 0.301 + 0.1708 = 0.4718$$

LD_{50} = antilog 0.4718

LD_{50} = 2.963%

Example 2

Dose (ppm)	Dead	Survived	Survived at this and greater dose Sg	Death at this and lower dose Ds	Sg + Ds	% Mortality $\frac{D_s}{S_g + D_s}$
2	1	9	27	1	28	4
4	3	7	18	4	22	18
8	4	6	11	8	19	42
16	6	4	5	14	19	74
32	9	1	1	23	24	96

It is evident from the data that a dose of 8 units resulted in a mortality of 42% and a dose of 16 units resulted in a mortality of 74%. The LD_{50} or the dose resulting in a mortality of 50% must be between 8 and 16 ppm.

The dose estimated to give 50% mortality can be calculated by the following formula :

$$\text{Log } LD_{50} = \text{Log } x_o + \frac{0.50 - P_o}{P - P_a} \text{ x } \Delta$$

Where Δ = minimum dose administered (in this case it is 2 ppm)

Log x_o = dose administered at the lower percent mortality than the required.

P_o = percent mortality at log xo

P = Percent mortality at higher dose than required.

$$\text{Log } LD_{50} = \text{Log } 8 + \frac{0.50 - 0.42}{0.74 - 0.42} \text{ x log } 2$$

$$\text{Log } LD_{50} = 0.90 + \frac{0.8}{0.32} \times 0.30$$

$$\text{Log } LD_{50} = 0.90 + \frac{24}{320}$$

$$\text{Log } LD_{50} = 0.90 + 0.75 = 0.975$$

$$LD_{50} = \text{antilog } 0.975 \text{ ppm}$$

$$LD_{50} = 9441 = 9.44 \text{ ppm}$$

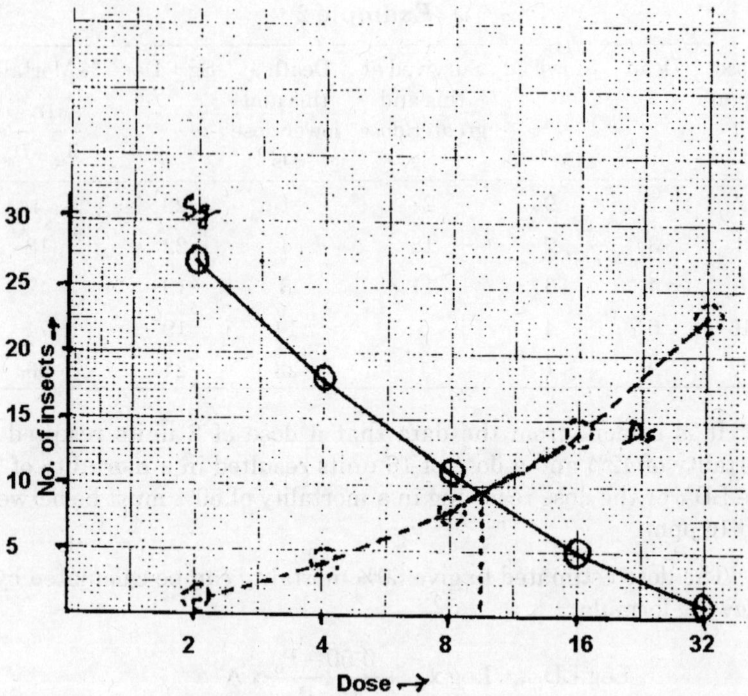

Fig. 6.32b.

ASSESSMENT OF LOSS

For estimating loss caused by pests, three samples of grains are drawn from the top, middle and lower portions of the bags in a godown. It is advised to collect samples from atleast 10% of the bags stacked. For example if 800 bags are stacked in a godown then take a random sample from 80 different bags. These samples are heaped. The heap is quartered and from this quarter a handful of grains are taken out. Suppose this handful grains counted were 500 and out of these 80 grains were holed and the rest 420 grains were apparently sound. Separate some 60 grains from these sound grains and cut them to see any infection or infecting stage like eggs larvae, pupa and adult. The percentage of loss is calculated in the manner given below.

500 grains contain 80 holed grains

1 grain will contain $\frac{80}{500}$

100 grain will contain = $\frac{80}{500} \times 100 = 16\%$

Now

if 60 grains contain 4 insects

then 420 grains will contain $\frac{4}{60}$ x 420 = 28 insects

since 500 grains contain 28 insects inside the grain

1 grain will contain $\frac{28}{500}$

100 grain will contain $\frac{28}{500} \times 100 = 5.6\%$

Therefore, the total damage in grain would be 16% + 5.6% = 21.6%

V. BRIEF ACCOUNT OF INDUSTRIAL ENTOMOLGY

Sericulture, Apiculture and Lac culture are popular agro-industries in India. A brief account of the equipments and steps in methodology which are used in these practices is given below.

SERICULTURE

Rearing equipments and accessories (Figs. 6.33a, b & c)

(i) Rearing stands

Rearing trays are placed on these stands. They are made of wood or bamboo.

(ii) Ant wells

The legs of the stands are kept in plastic/enamel bowls (circular or ractangular contaning water mixed with some insecticide so that the ants which are a serious menace to silk worms are not reached to the trays.

(iii) Rearing trays

They are made up of bamboo or other cheap wooden material. They are circular and used for rearing 1st and 2nd instars.

(iv) Paraffin paper

Paper coated with paraffin which is required to cover the trays to prevent humidity.

(v) Foam rubber

Soaked foam rubber is kept to maintain the humidity.

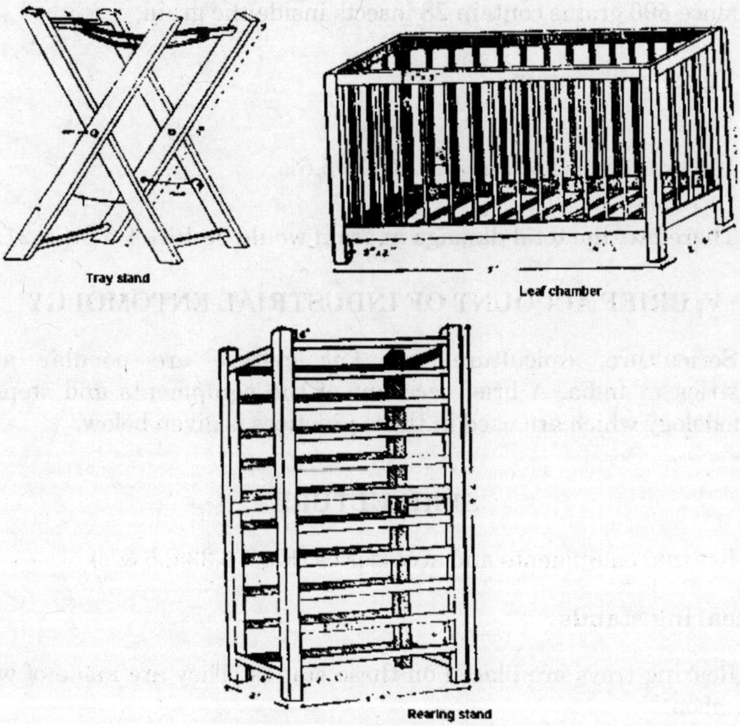

Fig. 6.33a. Rearing stand

(vi) Chopsticks

They are tapering bamboo rods which are meant to pick up younger stages of larvae to ensure their hygenic handling.

(vii) Feathers

They are used for brushing together newly hatched worms to prevent injuries.

(viii) Leaf chamber

Leaves are stored in chamber made up of wooden strips. It is covered with gunny bag to prevent the moisture during summers.

(ix) Chopping board and knives

Chopped leaves are offered to the worms. A small chopping board of wood and small and large sized knives are needed to cut the leaves for younger and older larvae.

(x) Cleaning nets

Since the manual separation of larvae from the litter has risk of injury and killing therefore nets made up of nylon (mesh size suitable for the different instars) are used for changing the rearing beds. Left over leaves and litter is filtered out.

(xi) Mountages (Chandrika)

Rectangular bamboo mat tied on four bamboo sticks bearing on it surface spiral bamboo tapes. The larvae which are ready to spin cocoons are transferred on to them.

Thermometer, Hygrometer are other minor equipments required to note the temperature and humidity of the rearing room.

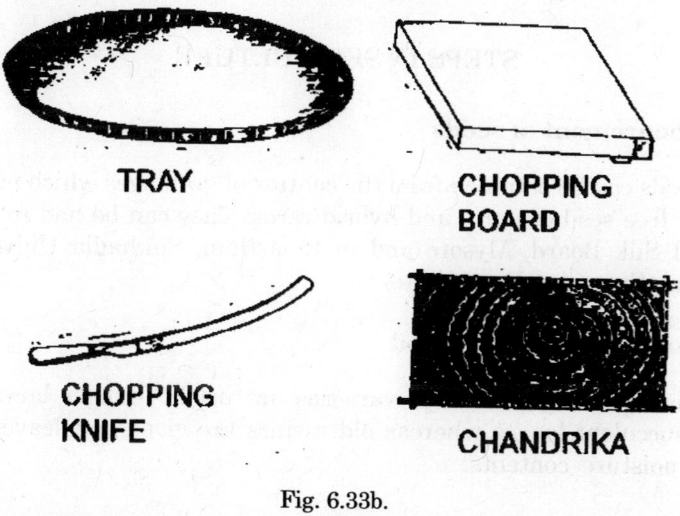

Fig. 6.33b.

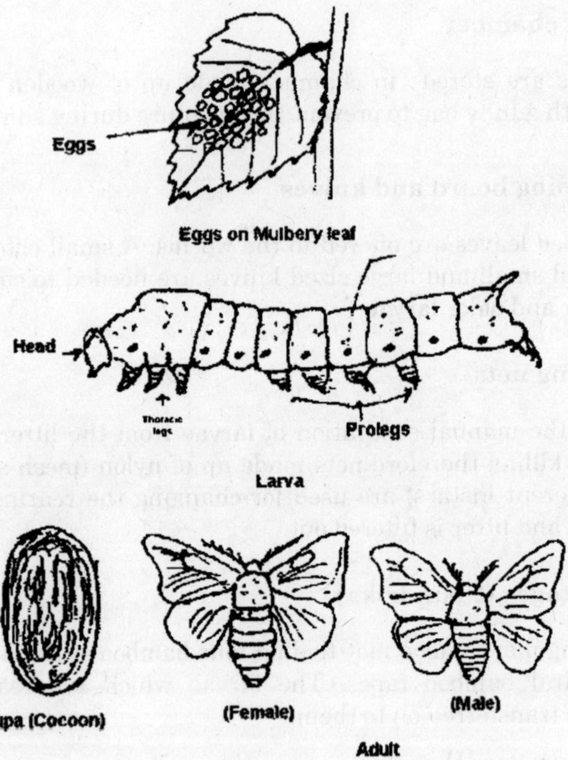

Fig. 6.33c. Life cycle of silk worm.

STEPS IN SERICULTURE

(A) Procurement of seeds

Seeds can be obtained from the centres of grainages which produce disease free seed of pure and hybrid races. They can be had from the Central Silk Board, Mysore and in Rajasthan, Sukhadia University, Udaipur (Extension Directorate)

(B) Provision of quality food

High yielding mulberry varieties are used. Younger larvae are given succulent leaves whereas old instars are given soft leaves with lesser moisture contents.

(C) Preparation of feed bed

Feed bed is a layer of chopped leaves spread on a tray. Four feeds a day are sufficient (9 am, 3 pm, 5 pm and 9 pm)

(D) Bed cleaning

Periodical removal of excreta and left over leaves is necessary. Net is commonly used. Net is spread on the old bed and freshly chopped food is given. The larvae migrate to the food. The old bed is cleaned.

(E) Spacing

Avoid overcrowding of caterpillars is called spacing. Leaves and worms are distributed with proper spacing. It can be done by increasing the size of the feeding trays.

(F) Moulting

Moulting is critical period. Bed cleaning at this stage is carried out. Small size leaves are given for pre moulting feeding. When the larvae settle down for moulting, feeding stops.

(G) Mounting

Transferring of mature fifth instar larvae on mountage is called mounting. The worm attach themselves to the mountage and starts spinning cocoon.

(H) Cocoon harvesting

It should be done at correct time. In tropical climate it is done on 5th day after their formation. It is normally done by hands. They are sorted out quality wise.

(I) Post harvest processing of cocoons

They include Stifling, Boiling, Brushing, Reeling and Re-reeling, Testing etc. The purpose of Stifling is to kill the pupae before they emerge into adults. It is done by steam stifling, hot air stifling and solar stifling, Boiling is done to make the cocoon thread reelable by dissolving the gum. Brushing is done to get the free end of the silk filament. Reeling is the process of unwinding the silk thread from the cocoon. It is re reeled on large reels. To judge the international standards, testing is done on size variations, quality, neatness etc.

APICULTURE

Bee keeping is gaining popularity now a days. It is done in artificial frame hives. Frame hives are fitted with movable frames on which the bees are persuaded to build their combs. They are usually composed of several boxes one on top of other in which are suspended hive frames. A movable frame hive, Langstroth is composed of the following parts.

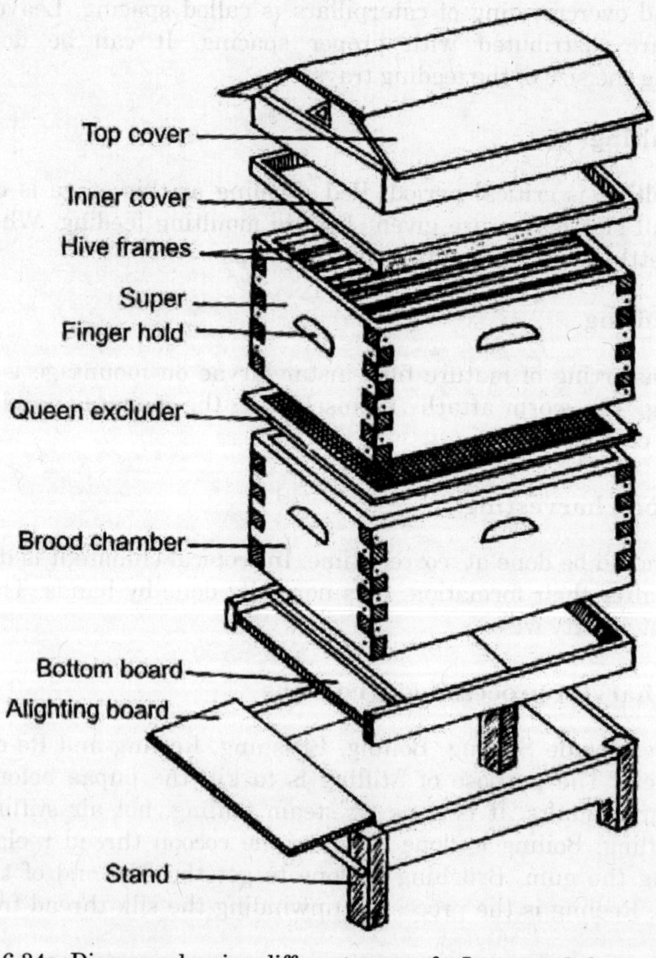

Fig. 6.34a. Diagram showing different parts of a Langstroth frame hive.

1. Stand

It is four legged structure with dimensions to support the floor board. It is 6-10 inch high from the surface of earth.

2. Floor board

It is a tray with all its four sides raised by side runners.

3. Brood box

It is a rectangular box. The number of brood box can be increased when the colony enlarges.

4. Hive frames

They are wooden frames. They are spaced apart to leave the bee space.

5. Queen excluder

It is necessary to separate the brood chamber from the supers where the honey is stored. This is accomplished by wire frame which allows the workers to pass through it but not the queen.

6. Supers

They are placed above the queen excluder. They are fitted with frames.

7. Covers

An inner and outer covers are there. Inner is flat board whereas the outer act as a roof for hive. Two oval holes are there in inner cover for bee escape.

The hives are painted white.

Other minor equipments used in bee keeping

Bee veil

It is made up of frames covered on four sides and a small mesh wire gauge. It is to protect the face of the bee keeper from stinging.

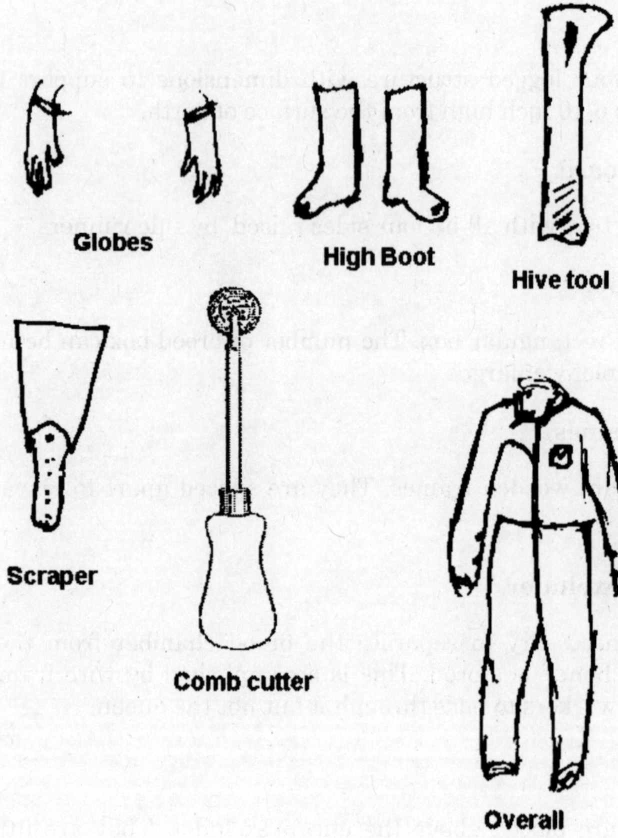

Fig. 6.34b. Some equipments used in Bee keeping

Overall

It is protective garment worn over the clothes.

Bee gloves

They are used as a protection to hands from bee stinging. They are of thick canvass or leather. They have elastic in wrist.

High boot

Boot which covers the ankle are high boots. They will prevent bees from entering into the trousers.

Hive tool

It is made up of steel. It is also used to scrap off the bee glue.

Scraper

One end is pointed and other bears a wooden handle. It is used for scraping propolis (bee glue)

Comb cutter

Old and worn out combs are cut and removed.

LAC-CULTURE

Lac is regarded as a subsidiary crop.

Harvesting

Cutting of lac encrustated twigs is called harvesting.

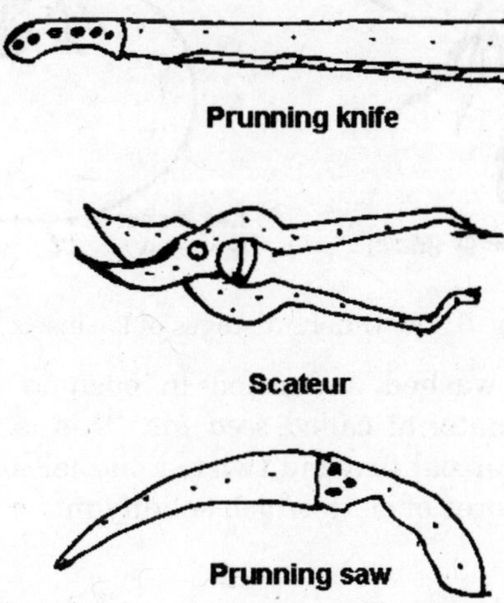

Fig. 6.35a. Some equipments used in Lac culture

Prunning

Cutting away old, weak and diseased branches. It is done with Scateur, Prunning saw and Prunning knife.

Lac extraction

When the nymph has escaped, it is stick lac that is left (also called phungi). These sticks are tied is bundles and immersed in water for 3-4 days by keeping them under heavy stones. Raw lac is now scraped while the stick is moist.

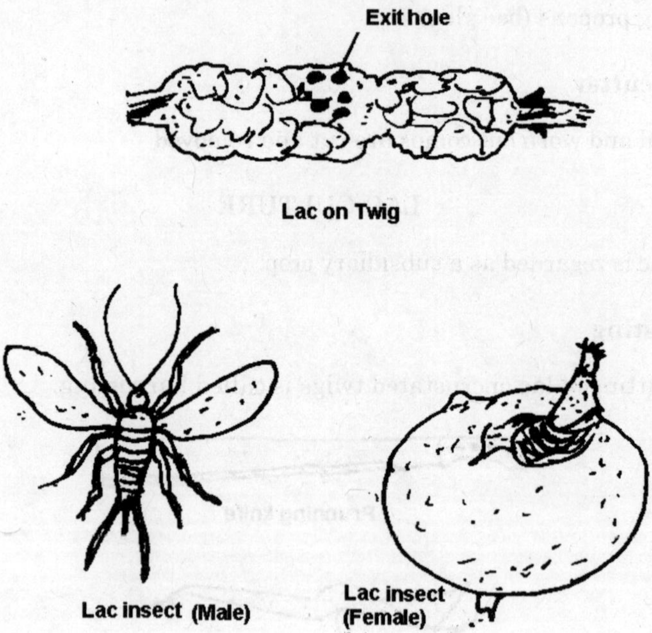

Fig. 6.35b. Different stages of lac insect

It is crushed, washed and dried in open to obtain on amber coloured granular material called *seed lac*. This is then packed into cloth bags before charcoal fire and twisted on melting. The molten lac flows out from the pores of cloth which solidify into round buttons. It is called *shellac*.

There are two strains - *Kusumi strain and Rangini strain*. Rangini thrives better on Palaas and Ber plants. Kusumi is grown on Kusum plant (*Schleichera oleon*).

CHAPTER - 7

EXERCISES BASED ON INSECT PHYSIOLOGY, ECOLOGY, TOXICOLOGY AND BEHAVIOUR

EXERCISE -1

Object - Identification of insect infestations in stored grains.

Requirements : 10% KOH solution, Beaker, watch glass, Petridish, Maize grains, Forceps etc.

Principle : KOH dissolves the food contents i.e. protein, carbohydrates etc., as a result of it the grains become transparent or opaque and any infestation there in can be detected.

Procedure : Take 10 grains of maize and soak them into 10% KOH solution in a beaker. Leave the grains in this solution for 24 hrs. Take out the grains with the help of forceps and observe.

Results : The grains become swollen and transparent or opaque and any infestations in the grain can be easily seen in the form of a dark hole on the grains and even the stage of development of the pest e.g. larvae, pupae, adult can be easily seen.

Precautions : Take care for not touching the 10% KOH with fingers as it is corrossive.

Inference : Out of 10 grains, the 4 grains were infested. The per cent infestation in the given sample is 40%.

EXERCISE -2

Object : To study the nerve cell in insects.

Requirements: House flies or Drosophila, Binocular, microscope (high power), Ether, Methylene blue (Saturated solution), Microsyringe/ Micropipette, Petridish, Coverslip, Slide, Forceps, Mineral oil and Brush.

Principle : Vital dyes like methylene blue on entering into circulatory system stains the neurons blue in colour.

Procedure : Narcotise the flies with ether vapours and put on the glass slide held with transparent tape exposing its ventral surface. Now inject the dye (methylene blue) in the thoracic region at the wing bases. Transfer the slide to refrigerator and take out after on hour or so. Now remove the wings and mount with cover slip in mineral oil. Observe under microscope. The axons are seen stained by blue colour. They appear as two concentric circles or dome shaped (Fig. 7.1).

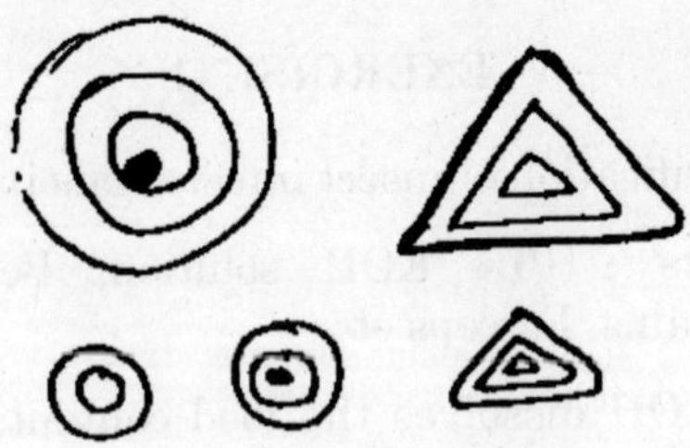

Fig. 7.1. Nerve cells

Results : In the proximity of campaniform sensillum the cell body is visible even with its dendrites. They are not only seen in wings but at wing bases and legs also.

Precautions (1) Methylene blue is vital dye. It only stains the living system of cells therefore the flies should be alive during the experiment. (2) Flies should not be dehydrated or moisture should not be lost.

Note : The nerve cells can also be stained by *Cabalt chloride*. Cut the ventral nerve cord of cockroach and put $CoCl_2$ solution. Cocl migrates through the nerves to the cell bodies. Now expose this nerve to the ammonium sulphide solution. The neurons and cell bodies exhibit black precipitation.

EXERCISE -3

Object : To study blood cells types in insects.

Requirements: Living cockroaches, Ringer's solution Microsyringe, Methylene blue and Eosin stain.

Principle : Methylene blue is a vital dye that stains the living tissues whereas eosin is a cytoplasmic stain.

Procedure : Put a drop of Insect Ringer's solution on a clean slide. Cut one antennae of cockroach and allow the stump to dip into the Ringer's solution. Shake the slide for few seconds and after a minute or so remove the slide. Now spread the contents (Ringer's solution + blood oozed from the cut antennae). Allow the cells to settle on the slide for few minutes and stain with methylene blue followed by eosin for differential staining (aquous). Dehydrate the slide in alcohol grade and mount in DPX or canada balsam. Observe the slide under the microscope. Different cell are seen.

Results : There are two types of cells in the haemolymph of cockroach viz. *amoebocytes* and *chromophils*. Amoebocytes are large cells with large nuclei and faintly stained whereas the chromophils are smaller cells with smaller nuclei but deeply stained (Fig. 7.2).

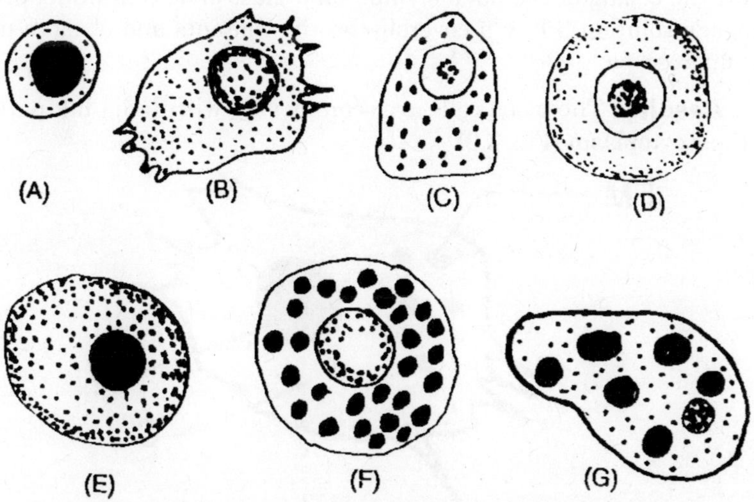

Fig. 7.2. Types of haemocytes (A) Prohaemocyte, (B) Plasmatocyte, (C) Granular, (D) Oenocytoid, (E) Cystocyte (F) Spherule cell (G) Adipohaemocyte (v) Coagulocyes or Cystocytes

Note : The haemocyte number varies in an insect. Free cells circulates in millions in cockroach. Prohaemocytes comprise more than 5% of the total haemocytes. Plasmatocytes and granulocytes accounts more than 30% each of the total haemocytes.

The number varies according to the stage of the insect e.g. 8000/ mm^3 in larva of Flesh fly and 12000/mm^3 in pupa. Generally the number in insects is about 40,000/mm^3.

Cystocytes (probably granulocytes) have not been recognised in Diptera, Hymoenoptera and Lepidoptera.

(**Suggested readings :** Insect haemocytes by A.P. Gupta (1979) Cambridge Univ. Press.)

EXERCISE -4

Object : To demonstrate the phagocytic activity (phagocytosis) in the insect cells.

Requirements : Living cockroach, Slide, Coverslip, India ink or China ink.

Procedure : Inject India ink into the abdomen of cockroach from its ventral surface. After atleast 40 minutes or so cut the antennae and allow the oozing of the haemolymph on a glass slide containing drop of Ringer's solution. Place a coverslip on the contents and observe under the microscope.

Results : The blood cells are seen moving with India ink particles in their cytoplasm (Fig. 7.3).

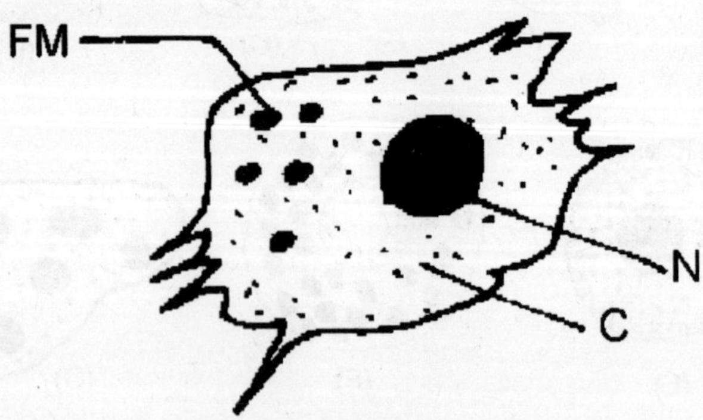

Fig. 7.3. Phagocytic activity
(N - Nucleus, C - Cytoplasm, FM - Phagocitized material)

Inference : Ink particles in the cytoplasm of amoebocytes and chromophils indicates their phagocytic activity.

EXERCISE -5

Object : To study the neuro endocrine control of metamorphosis in insects.

Requirements : Insect Ringer's solution, Housefly larvae, Dissecting binoculars, Petridish, Blotting paper, Needle, Forceps, Thread etc.

Principle : Moulting is mediated by the hormones secreted from the prothoracic gland which itself is activated by the brain. The neurosecretory cells in the brain secrete their secretions which causes ecdysial gland (=prothoracic gland) to release *Moulting hormone* (MH). Under the influence of this hormone, the larvae/nymphs moults into next stage i.e. pupae and adult.

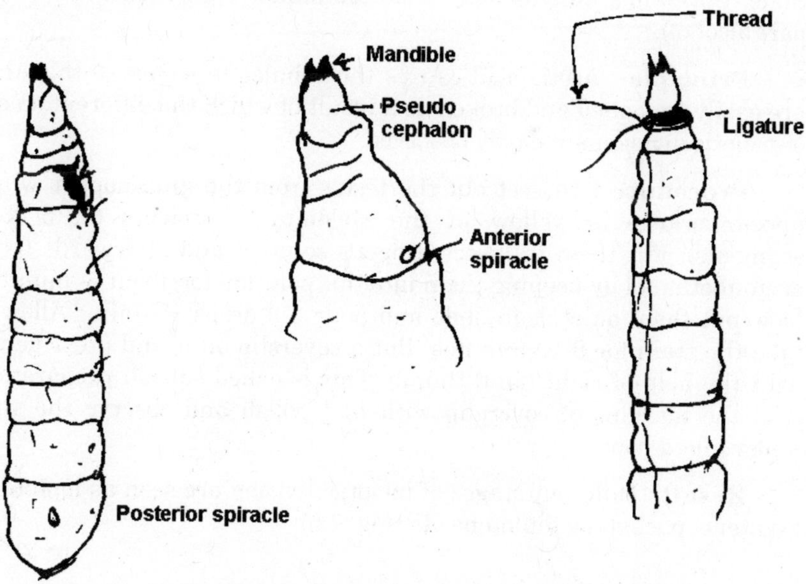

Fig. 7.4. Influence of Insect Hormones on Metamorphosis.

Procedure : Put the mature larvae on a slide in a drop of Ringer's solution. Immobilise the larvae with the help of ice cube and tightened the ligature at the junction of thorax and abdomen. Put these ligatured larvae on a damp blotting paper in a petridish and observe till the pupation takes place.

Results : Pupation takes place in the thorax region only (it becomes brown or darkened). The ligature has hinder the passage of hormone (MH) therefore, it does not pupate beyond ligature.

Inference : The *ecdysone* or *moulting hormone* secreted by the prothoracic gland is available in the thorax region therefore it pupates but the ligature did not allow this to percolate into abdominal region therefore it does not pupate (Fig. 7.4).

EXERCISE -6

Object : To study meiosis in grasshopper

Requirements : Grasshopper (male) or Cockroach (male) Insect Ringer's solution or physiological saline, Acetocarmine or Acetoorecin, Slide, Coverslips, Microscope, Aceto methanol (1 part acetic acid + 3 part alcohol).

Principle : Acetic acid causes the tubules to soften so that they are easily squashed and broken as a result of which the different stages of meiotic division are easily observed.

Procedure : Dissect out the testes from the grasshopper which appear orange or yellow in the abdominal segments (4, 5 & 6 segments). Put them in insect Ringer's solution and afterwards fix in aceto-methanol by keeping them into this solution for about 5- minutes. Now put them on a clean slide and pour 2-3 drops of stain. Allow to take the stain for 8-10 minutes. Put a coverslip on it and press gently with the help of right hand thumb. This is called squash preparation. Seal the margins of coverslip with nail polish and observe the slide under microscope.

Result : Different stages of meiotic division are seen as leptotene, zygotene, pachytene diplotene etc (Fig. 7.5).

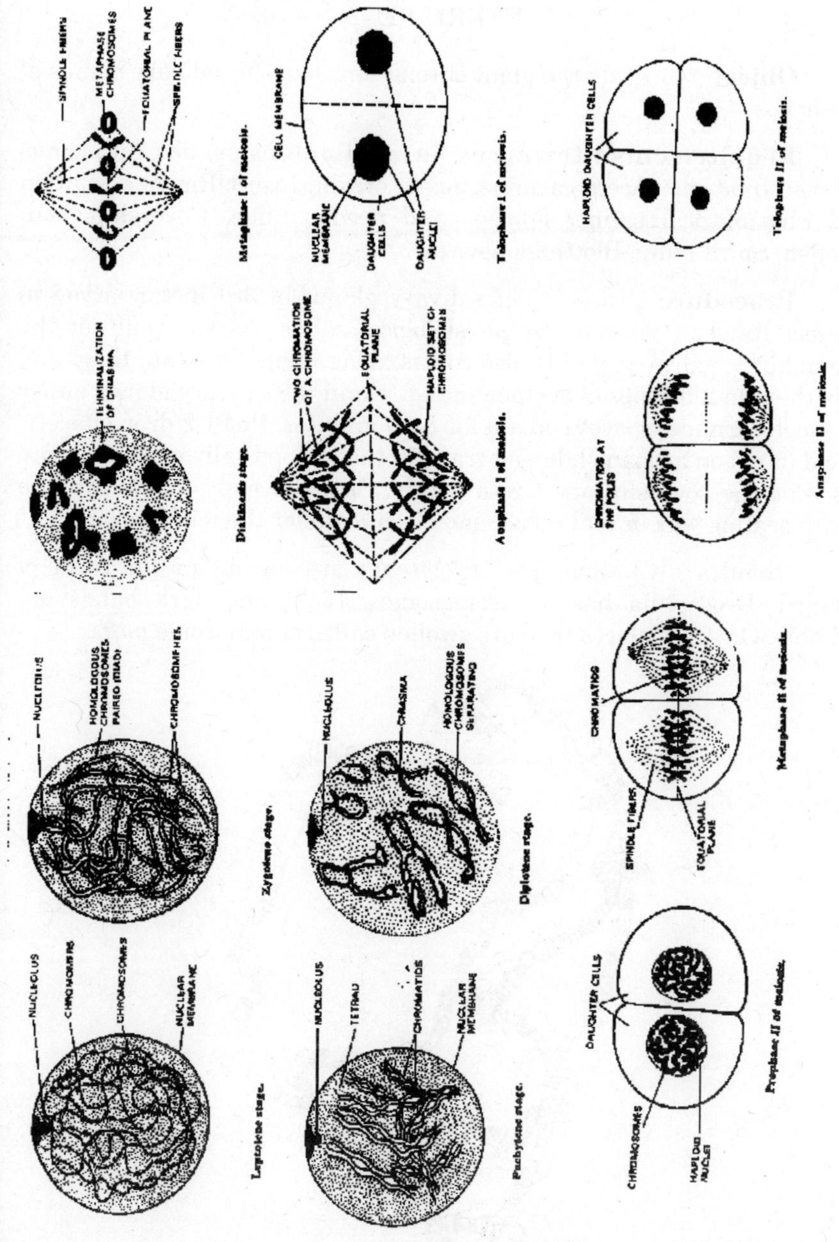

Fig. 7.5. Stages of Meiosis

EXERCISE -7

Object : To study the giant chromosomes in the salivary glands of insects.

Requirements : Drosophila larvae (third instar) or Chironomus larvae (mature), Aceto carmine, Aceto orcain, Insect Ringer's solution or Physiological saline, Forceps and needle, Slide, Covership, Nail polish, Spirit lamp, Blotting paper etc.

Procedure : Dissect out salivary glands of 3rd instar larvae in insect Ringer's solution or physiological saline. As you pull out the mandibles, salivary glands also comes along them. Separate them and fix them in 2-3 drops of acetomethanol. Stain the salivary glands either is aeotocarmine or aceto orcain for 8-10 minutes. Put 1-2 drop of acetic acid (50%) on a clean slide and transfer the stained salivary glands into it. Put the coverslip and press gently. Seal the margins of coverslip with sealing wax or nail polish and observe under the microsope.

Results : Chromosomes (Polytene) are visible in the form of bunch. Drosophila has 4 chromosomes. Light and dark bands are visible. On some places they are swollen called chromosome *puffs*.

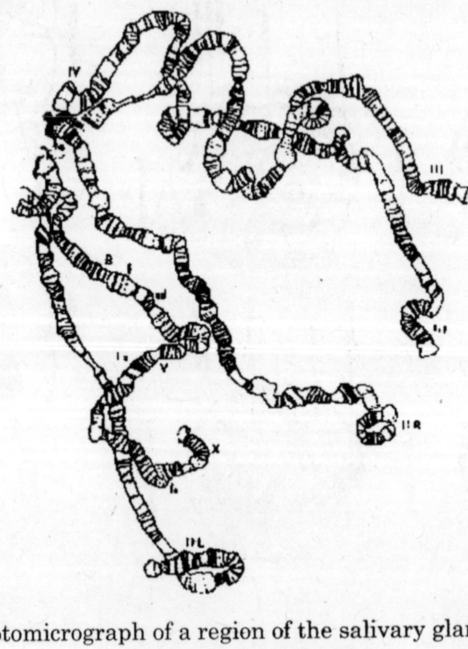

Fig. 7.6a. Photomicrograph of a region of the salivary gland chromosome *Drosophila melanogster.*

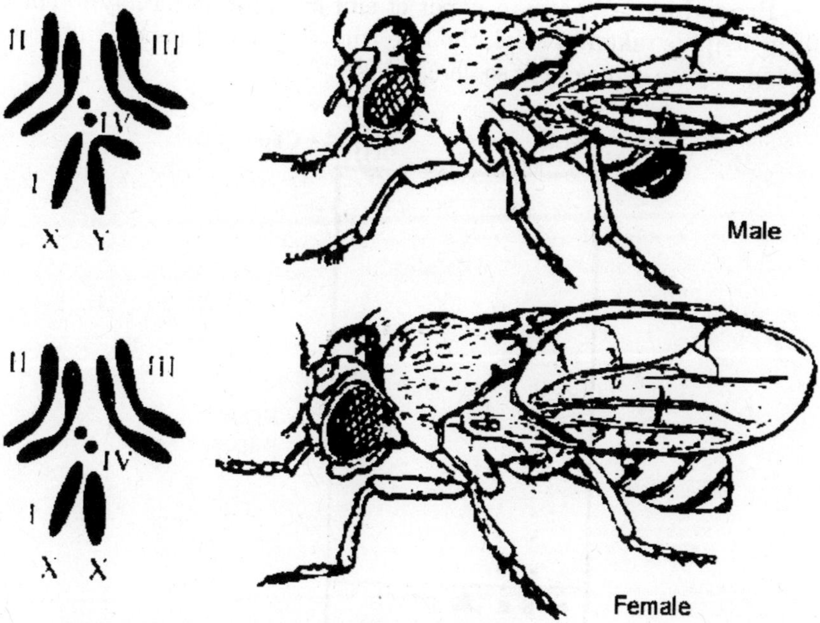

Fig. 7.6b. Male and female *Drosophila melanogaster* and their chromosomes.

EXERCISE - 8

Object : To demonstrate paper chromatography using insect haemolymph.

Requirements : Living cockroach or any other insects (grass-hopper, gryllus etc.), Whatman filter paper, Phenol, n-butanol, Acetic acid, Ninhydrin, Scale, Pencil, Capillary tube etc.

Principle : Low molecular weight components like amino acids sugars etc. can be separated by paper chromatography under the influence of a moving solvent through porous medium, the specific solutes can be separated. This separation depends on partition coefficient. When a small amount of solution containing a mixture of compounds is placed on the filter paper and a suitable solvent is allowed to flow, the substance migrate to a definite distance depending on its Rf value. It can be calculated by following formula.

$$\text{Rf (Ratio of front)} = \frac{\text{Distance travelled by solute}}{\text{Distance travelled by solvent}}$$

Procedure : Whatman paper of suitable size preferably (5 cm x 30 cm strip) is taken to mark the line with the help of pencil and scale leaving a space of 3.5 cm or so, lengthwise.

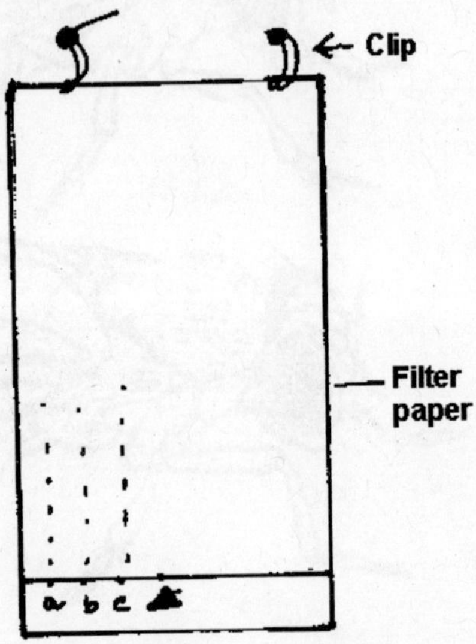

Fig. 7.7. Paper chromatography.

Mark spot a, b, c at equal distance and drop a solution of standard amino acid with the help of a capillary tube on mark `a' and `b'. At spot `c', put a drop of haemolymph and dip the paper in the solvent (80% phenol). (n-butanol, glacial acetic acid and water in 4:5:1 can also be taken). The solutes are allowed to move for 24 hrs. in an air tight chamber. When the solvent has run for the desired time remove it from the chamber and mark the solvent front with a pencil. Dry it in incubator or with the help of hair dryer and spray ninhydrin (locating agent). Calculate the **Rf** value and compare it with standard, Amino acids thus separated appeare as pink spots with ninhydrin. Encircle these spots with pencil and measure the distance. Calculate the Rf value.

(Rf value for a particular substance is constant in a particular solvent).

EXERCISE - 9

Object : To demonstrate the presence of chitin in the insect integument.

Requirment : Cockroach, KOH, Sulphuric acid and Iodine solution.

Principle : The presence of chitin can be detected by chitosan colour test. Concentrated alkali converts the chitin into chitosan which gives brown colour with iodine solution. Chitin is an important constituent of insect cuticle. It is a polysaccharide in the form of N-acetyl glucosamine in which the sugar residues are linked by 1-4 β-linkages. Adjacent chitin chains are held together by hydrogen bonds. About 50% of the dry weight of cuticle is chitin. It is tested by chitosan test or Van wisselingh's test. Chitosan test was demons-trated by Campbell (1929).

Procedure : Take a piece of tergum of cockroach and boil it in concentrated KOH. Take out the piece on a clean glass slide and wash several times in tap water. Now drop out one or two drop of iodine solution followed by a few drops of dilute sulphuric acid.

Result :On adding iodine solution the creamy translucent piece of tergum becomes brown. It turn violaceous when dilute sulphuric acid is added.

Precaution : Boil the KOH carefully till the piece of tergum becomes translucent.

EXERCISE -10

Object : To determine the moisture contents of insects

Requirements : Living cockroachs, Hygrometer, Culture jar (2 liter capacity), Stop watch.

Principle : Hygrometer has a device which reads the humidity directly on a graduated dial by a moving needle. During respiration the wet air is expired in the jar that can be directly measured.

Procedure : Take a clean and dry culture jar of 2 litre capacity and put the hygrometer into it. Cover the lid and stabilize the equipment for atleast half an hour. Note the reading of hygrometer (x_1). Now release 20 living cockroaches into the culture jar, cover it with lid. Note the reading of hygrometer after every 10 minutes. When there is no increase in the reading of hygrometer note down the final reading (x_2) and calculate the increase in humidity.

Calculation

n = number of living cockroaches or any other insect.

x_2 = moisture of in insects released during respiration.

Humidity before experiment - x_1 ----- (i)

Humidity after experiment - x_2 ------ (ii)

x_2-x_1 =xn (moisture of n insects)

EXERCISE -11

Object : To study the effect of para dichlorobenzene on Insects.

Requirements : Paradichloro benzene or Ethylene dichloride or Ethylene dibromide or CCl_4 or any fumigant, Petridish, Muslin cloth, Cotton, Petridish, Wool, Rubber band.

Principle : Fumigants are volatile chemicals whose vapours are toxic. These vapours on entering into the respiratory system, affect nervous system and cause death of insects.

Procedure : Place the cotton wool soaked in fumigant para dichlorobenzene in a petridish and cover it with muslin cloth. Now release a few stored grain pest-like Tribolium or Rhizopertha and release them on cloth. Now cover this petridish by another petridish. Observe the insects and note the time required for killing the insects.

Results : Insects show agility first showing cleaning movement, afterwards moves rapidly. When sufficient vapours are inhaled, the movement becomes sluggish and the insect dies. Ethylene bromide is more potential than that of para dichloro benzene (solid) or CCl_4 which are less volatile.

Precautions : 1. Ensure air tightness.

2. The vapours are toxic therefore precaution of not inhaling them should be taken.

Note : Fumigants are fast acting poisons. They are used against stored grain pests.

EXERCISE -12

Object : To determine LC_{50} of a pesticide/toxicant.

Requirements : Mosquito larvae, Any pesticide, Beakers, Semilogrithmic coordinate paper etc.

Principle : LC_{50} is that concentration of a toxicant or any pesticide that kills the 50% of the test population in a specified time . For example 24 hrs LC_{50} for a toxicant refers to that concentration of the toxicant at which 50% of the test insects are killed. LC refers to *Lethal concentration* and 50, to 50% population under test. Lethal concentration is used for aquatic medium only. LD_{50} is lethal dose and LT_{50} refers to lethal time.

Procedure : Take 6 beakers half filled with water and arrange them in a line. Mark I, II, III, IV, V, & IV. VIth one is control. Make desired concentration of pesticide of 0.5 ppm, 1 ppm, 1.5 ppm, 2 ppm and 2.5 ppm. Do not add any pesticide in "control" beaker. Release 10 larvae of mosquito of the same age group in beaker and observe the mortality for 48 hours. Record the result for 24 hours and 48 hours - respectively.

Observations

S.N.	Number of larvae	Concentration of a pesticide	Mortality during			
			24 hours		48 hours	
			Number	%age	Number	%age
1	10	0.5 ppm	nil	nil	1	10
2	10	1 ppm	3	30	4	40
3	10	1.5 ppm	4	40	6	60
4	10	2 ppm	7	70	9	90
5	10	2.5 ppm	10	100	10	100
6	10	Control	nil	nil	nil	nil

Note : They are completely hypothetical data. They are no where related with any pesticide, and are given just to understand the exercise and how to determine LC_{50}.

Result: Determination of LC_{50} by interpolation involves plotting the data on semi logrithmic coordinate graph paper showing concentration of toxicant on sub ordinate axis (y-axis) and percentage mortality data on abscissa (x-axis) (Fig. 7.8).

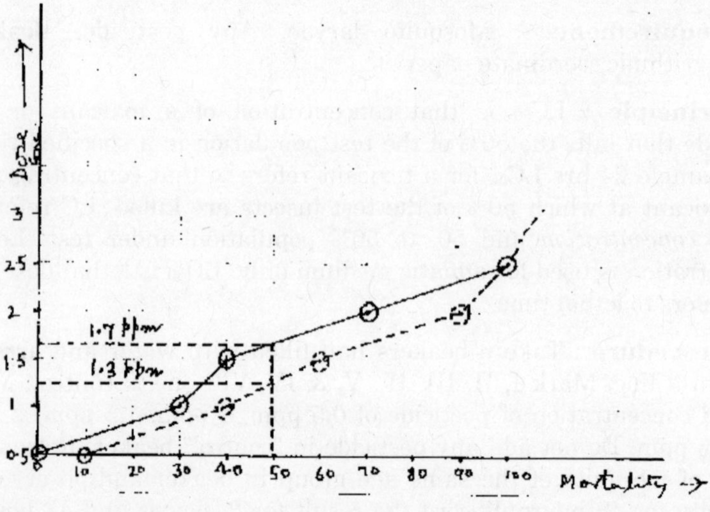

Fig. 7.8.

Thus plot the percentage of mortality observed in each replica. Draw a straight line between two points representing mortality for two successive concentrations. The concentration at which this line crosses 50% lethality line will be the LC_{50} value. This method is straight line interpolation.

LC_{50} for 24 hrs in this case is 1.7 ppm and for 48 hrs is 1.3 ppm.

EXERCISE 12(A)

Object : To demonstrate the occular effect of malathion (an organophosphate pesticide)

Requirements : Organophosphate pesticide (Malathion) Rat/ Rabbit or squirrel.

Procedure : Spray the pesticide from a distance and observe the test animals for eye inflammation, conjunctiva inflammation, lacrymation, tear discharge from the eyes. Such observations are made upto 24-48 hrs of exposure.

EXERCISE - 13

Object : To study the antifeedant and feeding deterrancy in insects.

Requirements : Biopesticides (Bioneem, Azadirachtin, Neem Oil, Salanin, Karanjin, Gedunin etc.) Stored grain pests like Tribolium, Callosobruchus, Rhizopertha, Sitophillus, Trogoderma etc.

Principle : Azadirachtin, isolated from the kernel of neem seeds is globally accepted antifeedant. Another triterpenoid salanin obtained from *Azadirachta indica* (Neem) is also a potent feeding deterrant to many insects. The active ingradient in "Bioneem" inhibits sensory receptors of mouthparts, resulting in disorientation of normal probing, feeding and intake of food.

Procedure : Take some known quantity of cereals and treat them with Bioneem or Azadirachin, or Salanin or Neem Oil (or other vegetable oils). Release the known number of stored grain pests like Rhizopertha or Trogoderma. In other container take untreated cereals of same quantity and release the same population of stored grain pests. After appropriate time (5-6 days later) record your observations on weight of the cereals before and after experiment, signs for number of insects feeding on grains etc.

Result : Stored grain pests which are extremely injurious to cereals could not feed as no feeding sign were there on the grains. The Bioneem did not allow the insects to feed as it is feeding deterrant and exhibit antifeedant activity. There was no loss of weight of cereals. Considerable mortality was also observed.

EXERCISE - 14

Object : To study the oviposition deterrancy in insects.

Requirements : Callosobruchus (Gram dhora) (Pest of stored pulses), pulses, vegetable oil (Mustard oil, Til oil, Groundnut oil), Neemoil or Neem extract.

Principle : Pulse beetle (female), recognise the surface texture of the grains for laying eggs. Oils makes the grain smooth therefore the marker substance released by female is not adhered to the grains due to greasiness created by oil therefore, it does not oviposit on such grains.

Procedure : Take some known quantity of pulse grains in a beaker and add some known quantity of any vegetable oils. Mix the oil to each and every grains. Now release 50 pulse beetles (sex ratio 1 : 1) and cover the container/beaker with muslin cloth tightly held by rubber band. In another container that serves as control, keep only the pulse grains (without vegetable oil) and release the same population of the pulse beetles. Observe the grains after 6-7 days and compare with control.

Result : No egg laying takes place in treated grains i.e. treated with vegetable oils whereas white dots appears on the pulse grains in control. Evidently the vegetable oil has exhibited the oviposition deterrency and the pulse beetle could not lay eggs on smooth surface of the grains created by oil.

BEHAVIOUR

EXERCISE -15

Object : To study thigmotaxis in insects.

Requirements : Earwigs (living), Houseflies, Ants, Beetles (Scarabeidae).

Principle : Response of an insect to touch is called *thigmotaxis*. When insects prefer to remain in contact with some surface all round its body is called *positive thigmotaxis* and when it prefers to remain in open is called *negative thigmotaxis*. Stimulus of touch tends to affect the behaviour of insects.

Procedure : Take a clean and dry beaker (capacity 500 ml) and release a few beetles. They tend to set at bottom and remains there. Repeat the exercise with earwigs, they also remains at the bottom and becomes immobilized. Now light on the lamp. Still they remains there otherwise earwigs run away from the light source and bury themselves in the soil. Repeat the exercise with ants and house flies also. They roam around the beaker and try to come out.

Result : The earwigs and beetles exhibit *positive thigmotoxis* and ants, and house flies *negative thigmotaxis*.

EXERCISE 16

Object : To study chemotactic response in insects.

Requirements : Drosophila flies, T-tube, Glass tubes, Fruits and Cotton.

Principle : Olfactory sense is developed in most of the insects. It helps in locating the food and recognition of the mate. The organs which detect chemical sense are located on antennae, mouthparts and legs.

Procedure : Take a `T' tube as shown in the Figure 7.9. Plug cotton to each open ends. Take 10 living Drosophila flies in base arm of `T' tube. Take out the cotton plugs from the arm of T tube inserting two glass tubes containing inert sand in one and fruit pulp in the other so that the Drosophila do not escape.

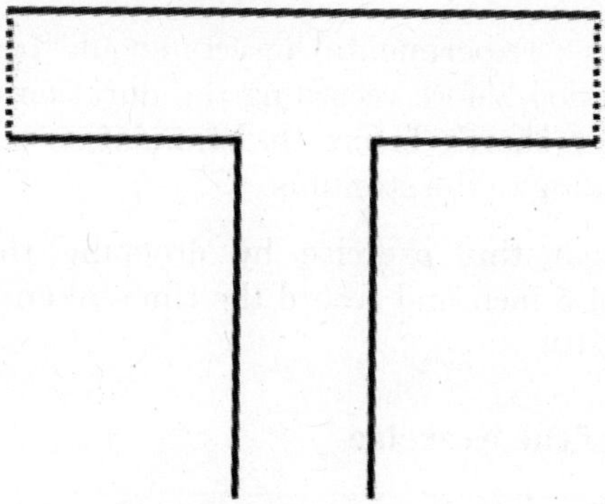

Fig. 7.9. T-tube.

Result : As the smell of the fruit pulp reaches to the flies in basal arm, they move towards the smell. Some flies also ventures in the tube containing inert sand but soon comes out of it and enters to the pulp containing tube.

Note : Carrion eating Beetles are also attracted to the smell of ammonia. The exercise can also be repeated with Dung beetle.

EXERCISE - 17

Object : To study Katelepsy in insects.

Requirements : Living insects (Beetles, Bug, Stick Insect), Stop watch, Glass rod, Beaker, Petridish, Muslin cloth, Rubber band, Brush, Forceps etc.

Principle : Various protective phenomenon are exhibited by insects. *Death feigning* is one of them and is displayed on mechanical stimulus or on approach of predator. Beetles and various other insects become motionless on touch stimuli, others roll into a ball (cuckoo wasp), others assumed bizzare postures and yet others becomes immobilized. This is commonly called *Katelapsy* or *thanatosis*.

Procedure : Take a Rhinoceros beetle and allow it to move normally. Now tap it with the help of glass rod and note the reaction of beetle. Also repeat the exercise with other beetles like *Holotrichia* and *Lasioderma.*

Result : On tapping it becomes motionless and pretend as if dead. It retract the legs and antennae. Note down the time of termination of thanatosis when the insect start moving its antennae.

Precautions : Experimental insects should remain undisturbed for reasonable period before recording the duration of thanatosis. Pre handling of insects shortly before the stimulation to `feign death' can prevent it responding to the stimulus.

Note : Repeat this exercise by dropping the beetle from a distance/height of 5 inch and record the time taken for termination of Katelapsy (Fig. 7.10).

Modifications of the exercise

(a) **Effect of mutilations on thanatosis** : Take 20 beetles of the same age group. Divide into 4 groups of 5 each and cut off antennae, head, legs (fore) and hind legs. Allow each batch to recover for 1-3/4 hour and induce thanatosis by tapping with rod or by dropping it from a height of 2-3 inches. Record the time taken for the termination of thanatosis in each group.

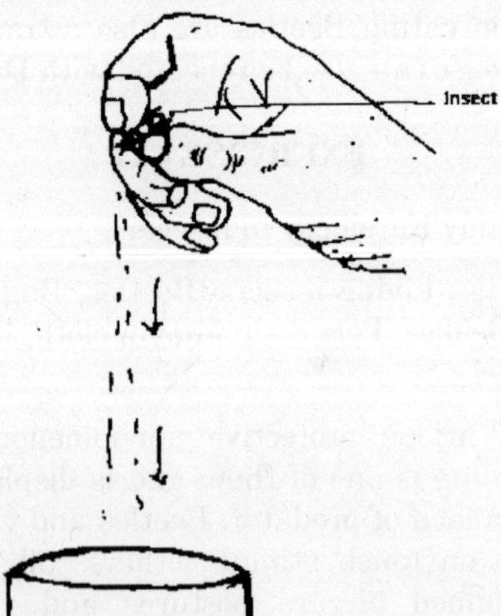

Fig. 7.10. Demonstration of thanatosis.

S.N.	Group	Part mutilated	Time taken for termination of thanatosis
1	I	antennae	—
2	II	head	—
3	III	fore legs	—
4	IV	hind legs	—

(b) **Effect of Light** : Take 5 beetles and induce thanatosis in dark and record the termination time.

EXERCISE - 18

Object : To study the antennal grooming in cockroach.

Requirement : Living cockroach, Glass trough, Glass rod, Acetone, Vaseline, Acetic acid.

Principle: Insects exhibit avoidance reaction to chemical `irritants'. The sensory organs for chemical irritants are "*basiconic pegs*" surrounded by *trichogen and tormogen cells* which are specially present on antennae. Each peg is associated with 4- 6 neurons whose dendrites reach to the tip. They carry the chemical sensations to the central nervous system and exhibit the behavioural response under the influence of brain. As per the directions from brain, the muscles found in scape (basal part of the antennae) contracts and bring the antennae to the mouthparts for grooming.

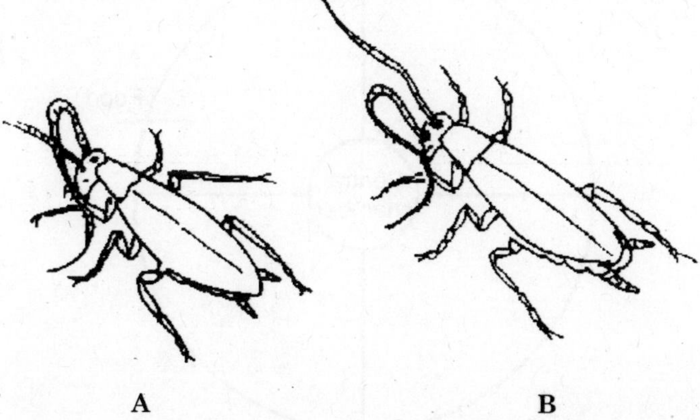

A B

Fig. 7.11. Antennal grooming (A. Grooming right, B. Grooming left antenna)

Procedure : Release a cockroach in a glass trough. Dip the glass rod in acetone and touch it to the antennae of cockroach. Repeat the exercise with vaseline and weak acid and record your results.

Result : When the acetone dipped rod is touched to the antennae, it immediately clean it with the help of mouth parts and legs.

EXERCISE - 19

Object : To study the food preference in Tribolium.

Requirements : Tribolium beetles, Wheat flour, Sooji, wheat bran, Maida, Glass container, Aspirator, Brush.

Principle : Stored grain pests attack a variety of food articles and grains but much of the damage is caused to that food which is more palatable and easily available. The insect restrict their feeding activity in most palatable and preferred food.

Procedure : Take a round plastic box or container which is equally divided into 4-5 chambers with a circular central chamber. Fill the chamber with different eatables viz. Wheat flour, Maida, Sooji Wheat bran and coarse grains. Release 50 Red rust flour beetles (Tribolium) in the central chamber. Cover the lid and allow the insects to select their most preferred food for 24 hours and record your results. Calculate standard deviation and standard error of the mean of your results.

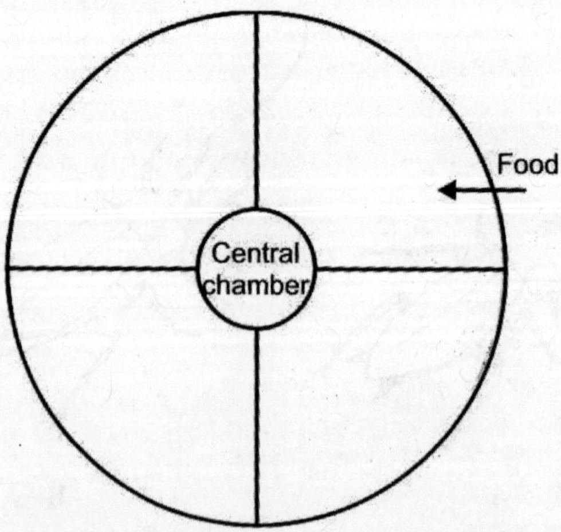

Fig. 7.12. Food Preference in *Tribolium*

Take your observations on the following parameters.

a. Number of insects in each food items.

b. Mortality if any.

c. Food stuff that record the lowest number of beetles.

d. Food stuffs that record the highest number of beetles.

Precautions: Central chamber should not be very deep otherwise the insects would not come out of it. If it is deep then make it shallow by pouring melted wax or plaster of paris (Fig. 7.12).

EXERCISE – 20

Object : To demonstrate the wing cleaning in Musca.

Requirements: Alive houseflies, Talcom powder, Brush, Vaseline, Beaker.

Principle : Response elicited to a particular stimulus constitute a particular behavioural activity which helps the insect in one way or other.

Procedure : Release 1 or 2 alive house flies in a beaker and cover it with muslin cloth. Now sprinkle some talcom powder on the wings of the fly. Repeat the exercise by applying vaseline on either of the wings and record your result.

Result : The fly immediately cleans the powder particles or vaseline by lowering the wings with the help of legs and it also jerks the wings.

EXERCISE - 21

Object : To study the response of light in House flies.

Requirements : Houseflies, Jar, Black Paper.

Principle : Many insects are attracted towards light, it is called *positive phototaxis* and those which are repelled from light are *negative phototactic*.

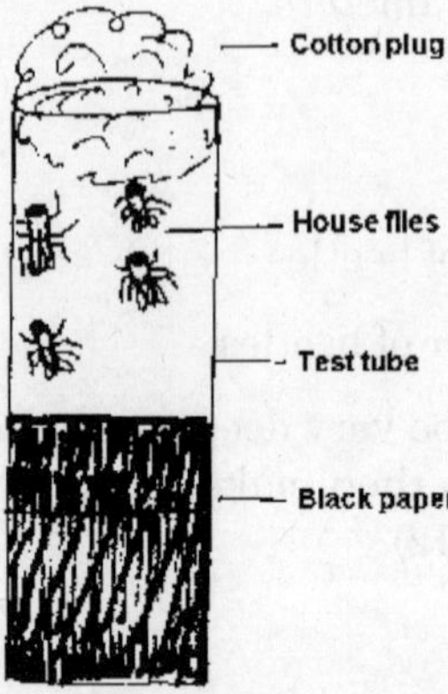

Fig. 7.13a.

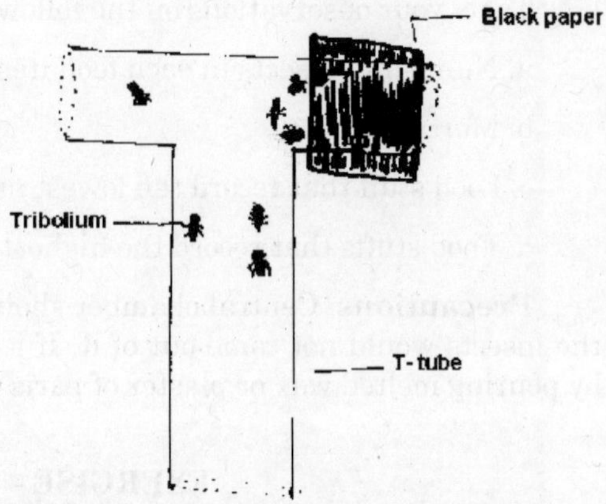

Fig. 7.13b. Showing light response in Tribolium. Arm of T tube is covered with black paper. Release some Tribolium in the central tube and note the response.

Procedure : Take a jar and wrap black paper in the middle part of it. Release some houseflies and note the resting of houseflies in dark and light part of the jar. Repeat the exercise with Tribolium.

Result : Majority of houseflies prefers to sit on the lighted part of the jar which indicates the positive phototactic movement.

EXERCISE - 22

Object : Demonstration of Biological control using Coccinella and Chryoperla.

Requirement : Petridish, whitepaper, Filter paper, camel hair brush etc.

Principle : In nature many insects feeds upon other insects which are injurious to our crops. Thus many crop pests are held in check by these carnivorous insects, the agent of biological control. Lady Bird Beetle and Chrysoperla are commonly found such agents of Biological control. They exercise control on aphids.

Procedure : Locate *Calotropis* plants in the fields and observe aphids in clusters on the leaves. Among the aphids the grubs of Lady Bird Beetle are found who feed on these soft bodied insects. The mustard aphids are also taken up by coccinella species. Collect the aphids in petridish or Beaker with the help of a Brush. Also collect some adult and larvae of coccinella. Set an experiment in the laboratory. Transfer 10 aphids to each petridish. and one grub of coccinella. Cover the petridish with another petridish and observe the grubs feeding on aphids. Note the number of aphids eaten by the coccinella grub and time in feeding. Calculate mean no. of aphids taken by the coccinella.

Subject the data to statistical calculations for standard deviation and Standard error of the mean. Interprete your results.

The experiment can be set with *Chrysoperla* as agent of Biological control and *Corcyra* eggs.

EXERCISE - 23

Object : To study seasonal abundance of Insects.

Requirements : Inch tape, Chord, Iron or wooden pegs, Notebook, Beakers, Petridishes, Brush, Paper envelops etc.

Principle : Abundance is described as the number of individuals per quadrat of occurrence.

Procedure : Lay quadrat of desired size (e.g. 1 m x 1m) at different locations and collect the total number of insects per quadrat (total and individual insects). The experiment can be repeated in different seasons. Identify the insects and interprete your results.

Calculations

A	B								C	D	E	F	G
	No. of insects and quadrat number								Total insects	No. of quadrats of occur-rence	Total quadrat studied	C/D abun-dance	C/E Density
	1	2	3	4	5	6	7	8					

$$\text{Abundance} = \frac{\text{Value in column C}}{\text{Value in column D}}$$

$$\%\ \text{Relative abundance} = \frac{\text{No. of species in a sampling unit}}{\text{Total no. of all species in that unit}} \times 100$$

EXERCISE - 24

Object : To calculate the growth index in Gram Pod Borer (*Helicoverpa armigera*)

Materials required : Glass vial, cotton Petridishes, Plastic vials.

Principle : Growth index can be calculated by dividing the per cent adult emergence by average developmental period.

$$\text{i.e. Growth index (GI)} = \frac{N}{Av}$$

where N = % adult emergence

Av = Average development period.

Procedure : Collect the adult moths at light trap and keep them in glass chimneys for 2-3 days. A cotton swab soaked in 10% sugar solution in kept in a petridish to nourish the adults. Folded paper strips are hanged for the egg laying. Eggs thus laid are kept at 30± 2°C temp. and 70 ± 5% R.H. The hatching and successive developmental stages are studied to calculate the growth index.

Calculations: Success index can be calculated by calculating larval period index, pupal period index, pupation index and adult emergence index.

$$\text{(A) Larval period index} = \frac{\text{Larval period in control}}{\text{Larval period in treatment}}$$

$$\text{(B) Pupal period index} = \frac{\text{Pupal period in control}}{\text{Pupal period in treatment}}$$

$$\text{(C) Pupation index} = \frac{\text{Percent pupation in treatment}}{\text{Percent pupation in control}}$$

$$\text{(D) Adult emergence} = \frac{\text{Percent adult emergence in treatment}}{\text{Percent adult emergence in control}}$$

$$\text{Success index (S.I.)} = \frac{A + B + C + D}{4}$$

EXERCISE - 25

Object : To study DNA by Feulgen reaction in salivary gland chromosomes.

Requirements - 1N-HCI, 5N-HCl, 10% Sodium metabisulphite, Feulgen stain, alcohol series. Tissue (salivary glands) of Drosophila. Pestel and mortar, graduated tubes, centrifuge machine, distilled water, Ice.

Feulgen Reagent

- Take 48 ml of 1N-HCl and mix it into 16 ml of double distilled water. Warm it at 60°C.
- Add to it, 0.8 gm Fuchsin slowly.
- Boil it (slowly) for 3 minutes.

Cool it (60°C) and filter. Allow to cool upto 30°C and then add 0.32 gm sodium metabisulphite. Stir, keep the bottle in dark for about 12 to 24 hrs. The colour is straw yellow. Now add 0.5 gm activated charcoal. Keep it for 12 hours in a cool place and filter through whatman filter paper. Keep in dark glass bottle in a refrigerator.

Modification : 0.5 gm basic fuchsin is dissolved in 100 ml of boiling distilled water and solution is cooled upto 58°C and filtered.

Temperature is allowed to fall upto 26°C and add to it, 10 ml of 1N HCl + 0.5 gm $K_2S_2O_5$ (Potassium metabisulphite). Keep the solution in a conical flask wrapped with black paper for overnight (24 hrs.). Magenta coloured solution turn straw colour (due to bleaching action of SO_2 released by the action of HCl on potassium metabisulphite). Add a bit of activated charcoal and keep it in a dark cool place and filter.

SO_2 solution : Take 20 ml 10% sodium metabisulphite and 20 ml 1N HCl. Add 360 ml of distilled water to it.

Principle - Acid hydrolysis removes the nucleotides and expose the aldehyde group. The exposed aldehyde group react with Schiff's reagent to give rise a purple colour.

Procedure

1. Dissect the salivary glands from 3rd instar larvae of Drosophila.

2. Fix the salivary glands in alcohol acetic acid (3 : 1) for 1 hours or so. Hydrate it by passing the glands in 95%, 70%, 50% and 30% alcohol series each for 15 minutes.
3. Place the hydrated slide bearing salivary glands into 5N HCl at room temperature for 20 minutes.
4. Rinse in two changes of distilled water.
5. Place the glands in Feulgen reagent for 30-90 minutes in a dark at room temperature.
6. Rinse it in 3 changes of 3 minutes each, in fresh SO_2 solution.
7. Dehydrate in 70%, 90% and 100% (two changes) of (4 minutes each).
8. Keep in xylene two changes and mount in DPX.

Result - DNA appears in shades of reddish purple.

(The intensity of the colour is proportional to the amount of DNA).

Second Method

Materials

1 N Hcl

Schiff's reagent

Sulphite wash

Fast green

For preparation of reagents see appendix 1

Procedure

(1) Fix the salivary glands of Drosophila or Chironomus or any other tissue (testes, ovary etc.) in carnoy's fixative for at least 20 minutes.

(2) Rinse in absolute alcohol for 3-5 minutes.

(3) Hydrate upto 70% alcohol.

(4) Rinse in distilled water.

(5) Incubate in 1 N HCl at 60°C temp. for 10 minutes.

(6) Again rinse in distilled water.

(7) Stain in Schiff's reagent for 10 minutes.

(8) Give 3 changes of sulphite wash of 2 minutes each.

(9) Wash in running tap water for 15 minutes (till sulphite smell is gone)

(10) Rinse in 70% alcohol.

(11) Counter stain in fast green solution for 4-5 seconds.

(12) Dehydrate in absolute alcohol.

(13) Clear in xylene.

(14) Mount in DPX.

Result - DNA appears as purple to reddish purple.

Third Method

Feulgen reaction is used for quantitative measurement of DNA. In Feulgen staining the nuclei are selectively stained and the intensity of the colour is directionally proportional to the amount of DNA present.

Materials required

Fixative - Alcohol acetic acid (3 : 1)

Bleaching solution : 10% Potassium metabisulphite + 5 ml HCl + 500 ml DW.

Tissue - Any insect tissue preferably testis and ovary.

Schiff's reagent - Take 0.5 gm. basic fuchsin in a conical flask. Add to it 100 ml boiling water.

Mix thoroughly and allow to cool to 50°C. Add 0.5 gm potassium metabisulphite and 10 ml 1N HCl.

(1N HCl = 10 ml conc. HCl + 90 ml. DW)

keep in air tight bottle for a day. Add 0.25 gm activated charcoal and filter it. If it is not colourless further add activated charcoal and keep in refrigerator.

Procedure

1. Fix the tissue in alcohol acetic acid for 3-10 hours.

2. Transfer to 70% alcohol. Dehydrate in 90% alcohol, absolute to xylene.
3. Embed the tissue in wax (paraffin).
4. Make blocks and cut the section at microtome (5.8 μ thick).
5. Remove the wax (deparaffinise) from the tissue by passing through descending series of alcohol.
6. Bring the slide to 1N HCl for 5-6 minutes at 60°C temp.
7. Bring it to 1N HCl room temperature.
8. Rinse in distilled water.
9. Stains in Schiff's reagent for 1-2 hrs.
10. Drain and transfer through 3 changes of bleaching solution (5 ml 10% of pot. metabisulphite + 5 ml 1 N HCl + 90 ml distilled water) for 1 minute each.
11. Rinse in distilled water.
12. Dehydrate upto 90% alcohol.
13. Stain in 0.005% Fast green in 95% alcohol.
14. Dehydrate in absolute alcohol. Clear in xylene and mount in DPX.

Result - The nuclei stain purple representing the DNA.

EXERCISE - 26

Object - To study the number of instars in insects. *OR* To determine the developmental stages in insects. *OR*. Verification of Dyar's law.

Requirements - Culture of hemimetabolous and holometabolous insects like bugs, butterflies etc. Scale, Graph, Pencil paper.

Principle : Dyar's law states that the width of the head of a larva in successive instars show a geometric progression in their increase. By using such criteria, it is possible to ascertain the number and stage of instars.

Procedure : Measure the width of the head capsule of larva by placing it on the graph paper.

For example - Following hypothetical data represents following results.

Head width of the larva - 0.75 mm 1.0 cm, 1.5, 2.0 and 3.2 cm.

Divide 2.0 by 3.2 (It will give the ratio of increase)

This ratio is used as factor = 0.625

now calculated the width for

Ist instar	- 1.0 x 0.625	= 0.625
2nd instar	- 1.5 x 0.625	= 0.937
3rd instar	- 2.0 x 0.625	= 1.25
4th instar	- 3.2 x 0.625	= 2.0

	I	II	III	IV	V
Observed value	0.75	1.0	1.5	2.0	3.2
Calculated value	0.625	.937	1.25	2.0	....

Result - Such studies are useful in predicting the probable number of instars thus, help in the study of bionomics.

EXERCISE - 27

Object - To study avoiding reaction in insects.

Requirements - Aquaria or Glass trough, wax, Aquatic Beetles.

Principle - The movements are perceived by the organs located in the second segment of antennae, probably the *Johnston's organs.* If the antennae are cut off, the aquatic beetles collide while swimming.

Procedure : Release few water beetles in the aquarium and observe their movements. While swimming they never collide even after the walls of the aquarium are darkened. Now remove all same from the bottom of the aquarium and coat the sides of the aquarium with paraffin-wax at the water level so that the meniscus would disappear.

Result : Observe the beetles, they start colliding while they swim.

Inference : Beetles see the movement of particles at the surface by the reflection from the wall of aquarium or vibrations of their own movements, thus avoid colliding each other. If the antennae are cut off from the base the ability to avoid collision also reduces and disappears.

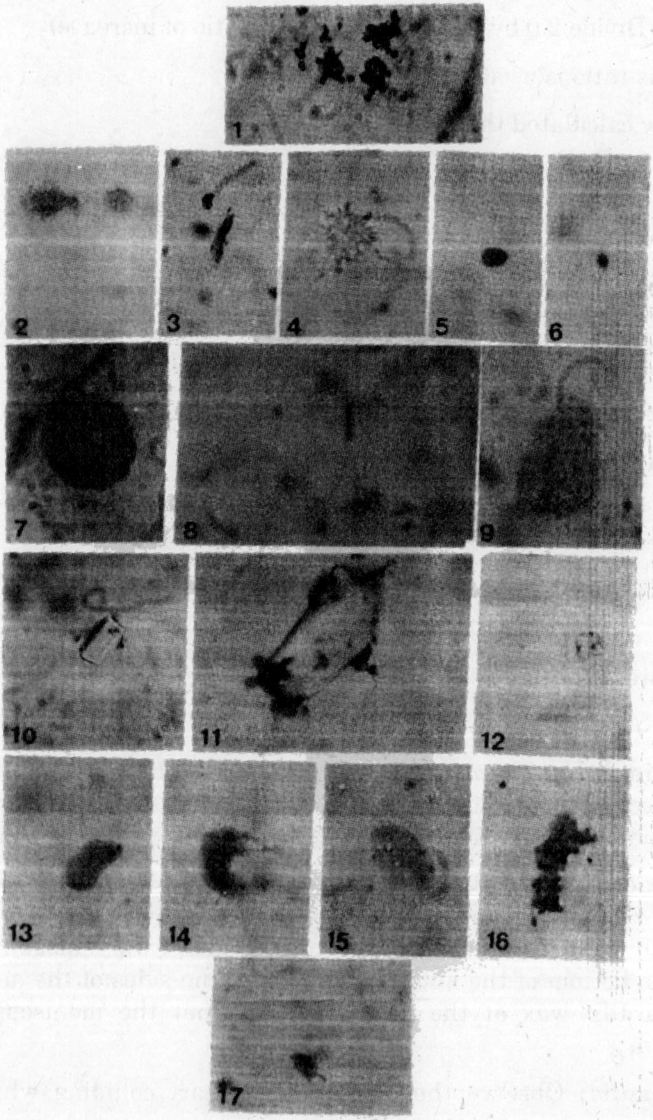

Photograph: **Fig. 1.** Prohaemocytes (PR) x 400; **2-6.** Variants of Plasmatocytes (PL) x 250. **7.** Granuocytes (GR) x 400; **8.** Granuocytes (GR) x 250; **9.** Granuocytes (GR) x 400; **10-12.** Variants of Coagulocytes (CO) x 400;

13. Oenocytoid (OE) x 400, **14.** A crescent shaped OE x 400; **15.** An adiphoaemocyte (AD) x 400; **16.** A Spherulocyte (SP) x 400; **17.** A Podocyte (PO) x 250 (Courtesy: S.C. Pathak and Sunita Srivastava, 2007)

EXERCISE - 28

Object: Preparation of Blood Film of Brown banded cockroach (*Supella* sp.).

Requirements: Lambda pipette, 5% formaldehyde, glass slide, Giesma.

Procedure: Collected the haemolymph blood from the cut leg of the insect with the help of lambda pipette. It is then transferred to a drop of liquid fixative (5% formaldehyde) on a clean slide. Spread the drop evenly with the help of a cover slip and dry it in air. Stain with Giesma solution observe under the microscope light/phase contast.

Results: Eight types of haemocytes can be identified (from the nymph and adults). Prohaemocytes, plasmatocytes, granulocytes, coagulocytes oenocytoids, adipohaemocytes, spherulocytes and podocytes.

CHAPTER - 8

INSECT ANATOMY

I. PREPARATION OF PERMANENT SLIDES

1. D.P.X. mounts

Permanent slides of different parts of the insect body or tissues or whole mounts of minute insects can be prepared in the Laboratory by dehydrating the material in different alcohol grades. Follow the steps given below :

Bleaching (10% KOH) if necessary

(15-20 min)

Staining (2% aquous mercuro chrome or acid fuchsin)

(20 min)

70% alcohol

(20 min)

90% alcohol

(20 min)

Clove oil

(1-2 min)

Mounting (DPX or Canada balsam)

(Preserved tissues can directly be transferred in 70% alcohol. Time allowed for dehydration at each steps depend on the size of the tissue. Large tissues and hard structures takes more time. Hard and dark structures are kept in 10% KOH till they become colourless and translucent).

(i) Bleaching

Heavily pigmented forms can be bleached in chlorox for a few minutes then 10% KOH solution. If the insect is dry, put into 90% ethyl alcohol to wet it and then taken through 10% KOH and distilled water.

(ii) Staining

If it is necessary to stain then put the tissue in acid fuchsin (aquous) for 15-20 minutes. Transfer to distilled water till the deep colour is diffused. Put it into 70% alcohol. 2% Mercuro-chrome (aquous) can also be used. Dilute the stain before use by adding drops of distilled water 1 : 2 (one part stain; 2 parts distilled water).

(iii) Dehydration

70% alcohol is the starting point for bleached tissues. Leave the tissue for 20 minutes in it then transfer to 90% alcohol for the same length of time. Unbleached specimens or tissues can directly be put into this grade of alochol.

(iv) Clearning

Clove oil is a good clearing agent. Leave the tissue in clove oil or eugenol till it becomes transparent. It now sinks down. The tissue can also left for indefinite time in this step.

(v) Mounting

Locate the centre of the glass slide and drop out canada balsam or DPX on it. Spread it and transfer the structure or specimen from the clove oil. Lower the cover slip with the help of needle and forceps. Orient and arrange the specimen by putting additional drop of DPX if required. If the specimen is thick then use small squares of broken glass or coverslip or celluloid.

(vi) Drying and Labelling

Slide thus prepared is left for drying atleast for two weeks. It can also be put into the incubator or oven at 45-50°C. Paste a label with basic informations naming the structure or part, date and name etc.

2. Lactophenol mounts

Apterygotes which are small and delicate are best mounted in this material. They can be mounted directly from alive forms or preserved forms in 70% alcohol or other grades.

Put a small drop of the medium (polyvinyl alcohol + Lacto-phenol) on a slide. Transfer the specimen either alive or preserved, orient and arrange the specimen. Use glass pieces or celluloid support if the tissue is large and thick. Put a drop of the medium and lowers the cover slip gently. Air bubbles can be driven out by gentle warming. Dry it in oven for a day or two or 2-3 weeks at room temperature.

Note: Prepare a permanent slide of Protura, Diplura, Collembola, Thysanura, Anoplura, Mallophaga, Thysanoptera, Mosquito larvae etc. See the Appendix-I for preparation of polyvinyl alcohol-Lactophenol medium.

Mounting of Aphids, Barklices and Book lices

I. Temporary mounts

Faure, Berlese or chloral hydrate-gum arabic mixtures are used for mounting living or fresh specimens directly. The slide can be placed in an oven at 120°F for 60 minutes to permit clearing and inflation. Suitable sealing material can be used for sealing the slide e.g. muryite.

II. Permanent mounts

Two methods are recommended

a. Caustic Potash

1. Fresh material or preserved or dried is immersed into 10% KOH or NaOH
2. Keep in a warming oven, boiled over a flame or set aside in close containers.

3. Transfer it to acetic acid, tease or press out KOH/NaOH and remove it.
4. Immersed into fresh acetic acid for atleast 10 minutes.
5. Stain with acid fuchsin or alcoholic fast green, if necessary.
6. Keep in clove oil for 5 minutes.
7. Transfer directly to D.P.X. or Canada balsam or Euparal. Cover with coverslip.

b. Lactic acid

1. Transfer the fresh specimens or preserved into the Lactic acid (50 parts) + alcohol (35 parts) + water (15 parts).
2. Keep in oven at 120°F for 24 hours (fresh specimens) and 48 hours (preserved specimens).
3. Tease out the embryo stain if required.
4. Transfer to acetic acid then to clove oil.
5. Mount in DPX/Canada balsam/Euparal OR transfer to 95% alcohol then to absolute alcohol then clove oil and mount in Canada balsam/DPX/Euparal.

Mounting of Thrips

1. Keep the fresh specimens into 5 or 10% KOH or NaOH solution or Lactic acid. Dehydrate it in alcohol, series clear in clove oil/cedar oil/xylol and mount directly into canada balsam/DPX/ Euparal. Very good mounts may also be prepared by placing the material to Berlese, Faure or similar media heating to 100°F for 60 minutes. Ring the slide with suitable cementing material staining (if necessary) can be done in acid fuchsin or fast green.

Permanent slide of Mouthparts

Dissect out the mouth parts from different insects and make a permanent slide and study the different structures and sub structures.

— **Biting and chewing types** (Cockroach, Locust, Grasshopper and Beetles)

— **Piercing and sucking type** (Mosquito and Bugs)

— **Chewing and lapping type** (Honey bee)

— **Sponging type** (Housefly)

— **Siphoning type** (Butterfly, Moth)

Chewing and Biting type

Example : Cockroach and grasshoppers

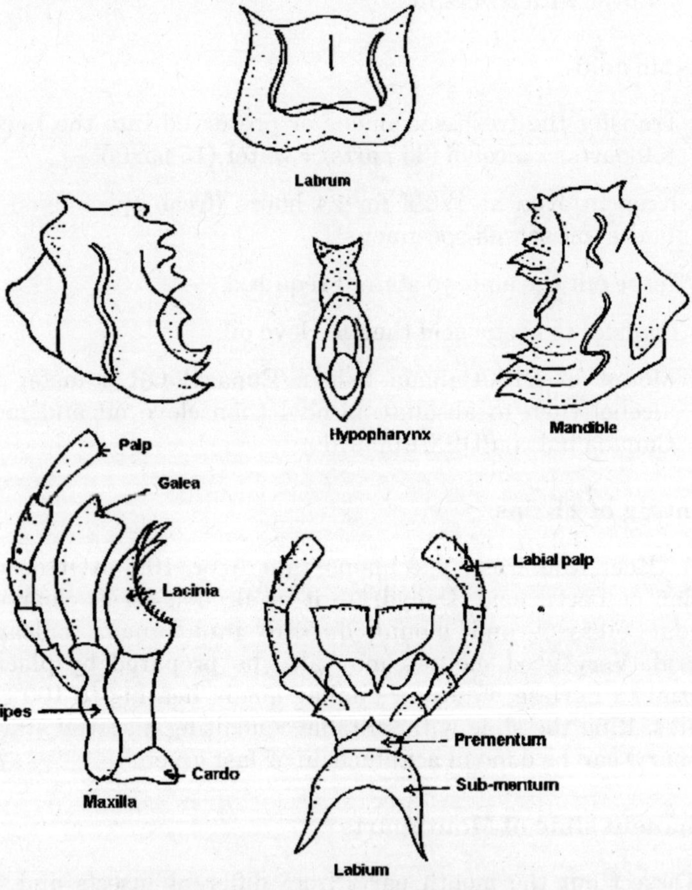

Fig. 8.1. Chewing and Biting type.

1. Two pair of mandibles/jaws present.
2. Maxilla and mandibles works together for tearing off and masticating the food.

3. Jointed palps are present.

4. Mandibles cut off and grind the food. Maxillae and labrum push it into the oesophagus.

5. Labrum and Labium present.

6. Locust, Crickets, Earwigs, Weevils, Lepidopteran larvae possess such types of mouthparts.

Note: *Grasping* present in soldier ants and stag beetles and *grinding* present in caterpillars, beetles and orthoptera are modifications of chewing type.

Piercing and sucking type

Example : Mosquito and bugs

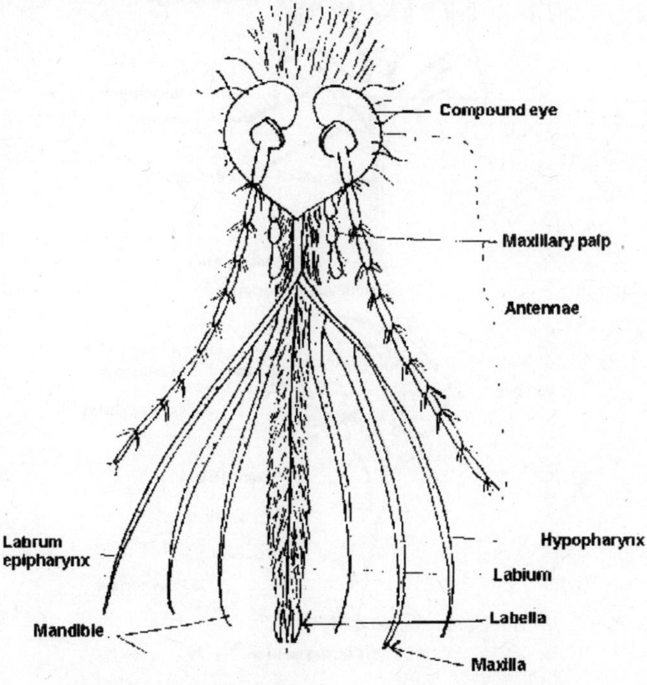

Fig. 8.2. Mouthparts: piercing and sucking type

1. Mouthparts are elongated and slender to form a beak like structure.

2. Needle like styles are enclosed and protected with lower lip.

3. Maxillae and mandibles are slender and elongated.
4. Upper lip is short and flat.
5. Hypopharynx is elongated.
6. Palps are either absent or poorly developed.
7. All these stylets are joined to form a food channel and salivary duct.
8. Stylets are forced into the food source or tissue, it is sucked up by pump formed by the wells of mouth cavity.
9. Bugs (Heteroptera) Lice (Anoplura), fleas (Siphonaptera) and Mosquitoes (Diptera) posses such type of mouth parts.

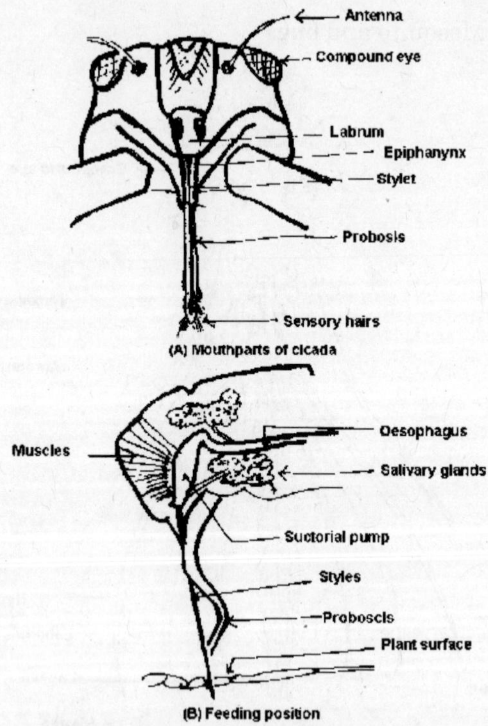

Fig. 8.2(b). A mouthparts of Cicada

Rasping and sucking type

Example : Thrips

1. Rasping cone, made by stylets, is present.
2. Left mandible, maxillae and hypopharynx is elongated and appears stylet like.
3. Right mandible is reduced.
4. Both pair of palps are present.
5. Upper lip (Labrum) and hypopharynx constitute the food channel that lies in the head capsule and lower lip along with hypopharynx forms salivary duct. Styles do not enter deep into the wound but the sap exuded is sucked up by the rasping cone.
6. They are present in thrips.

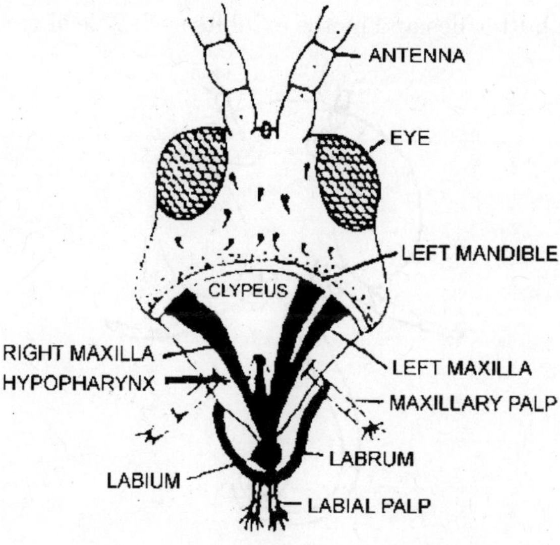

Fig. 8.3. Rasping and sucking type (Thrips)

Siphoning type

Example : Moth and Butterflies.

1. Such mouthparts are highly specialized for sucking the liquid food or nectar.
2. Upper lip is much reduced.
3. Mandibles usually absent.

4. Palps are rudimentary.
5. Lowerlip consist of two parts, a broad plate and three segmented labial palps.
6. The galea of maxillae join together to form a slender hollow tube like proboscis which remains curled under head when not in use.
7. Feeding involves the uncoiling and coiling of proboscis. The coiling is affected by many tiny muscles present inside the proboscis whereas the uncoiling is initiated by the pressure of haemolymph forced into the tube.
8. In the proboscis runs a food channel through its full length and the food is sucked up through its food channel.
9. Adult butterflies and moths exhibits such type of mouthparts.

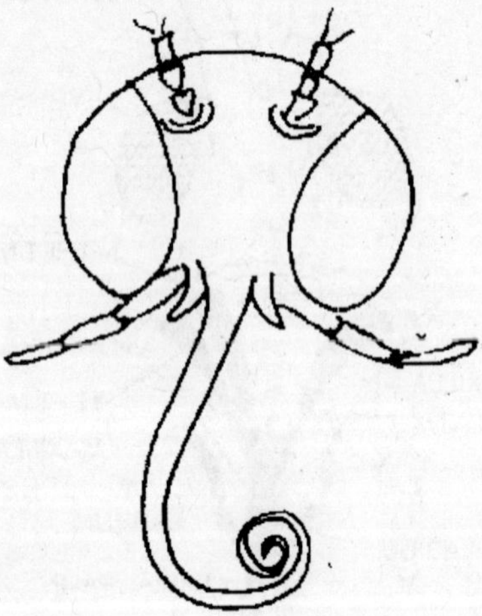

Fig. 8.4. Siphoning type.

Chewing and lapping type

Example : Honey bee

1. Upper lip and mandibles are of chewing type and helps in grasping the food. Maxilla and labium are elongated.
2. Palps are small.

3. Galea with its concave inner surface forms a roof over glossa and fits length wise against the labial palps forming a food channel.
4. Mandibles and upper lip helps in grasping the prey.
5. Glossae are fused to form a channeled organ called alagossae.
6. The nectar is sucked up by the capillary action of glossa. Shortening of glossa takes place due to pull by muscles as a result the nectar is squeezed and deposited at the base of paraglossa.
7. Bees, wasps ants and other Hymenopteran insects possess such mouthparts.

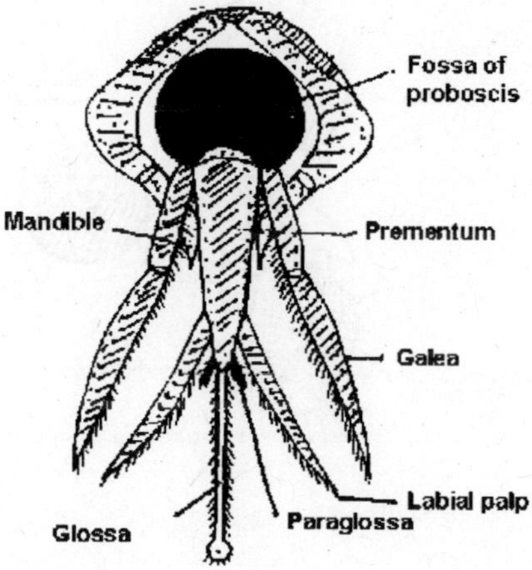

Fig. 8.5. Chewing laping mouth parts of a honeybee

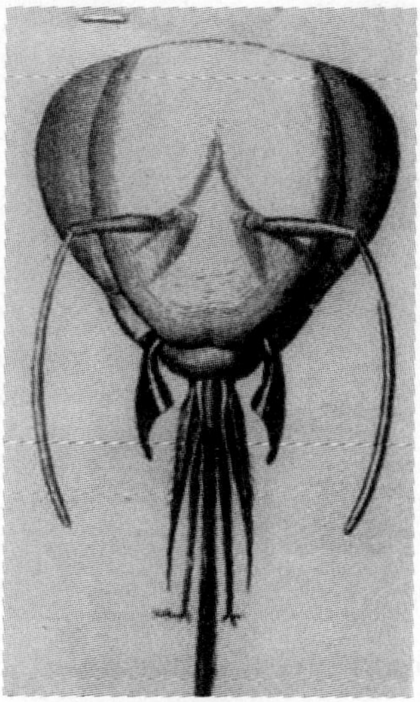

Photographs. Head and mouthparts of Honey bee.

Sponging type (Housefly)

1. Mandibles are absent.
2. Maxillary palps represents the maxilla.
3. Lower lip is divided into rostrum and houstellum, the end of labium is modified into sponge like structure labella. Labella is traverse by pseudotracheae which converge at one point and open into the food channel.

4. Hypopharynx and epipharynx constitutes food channel which leads into oesophagus.
5. Liquid or readily soluble food in saliva are taken. The labella is thrust to liquid food that is taken up by capillary action of pseudotracheae. The food is then passes to oesophagus.
6. Houseflies exhibits such type of mouthparts.

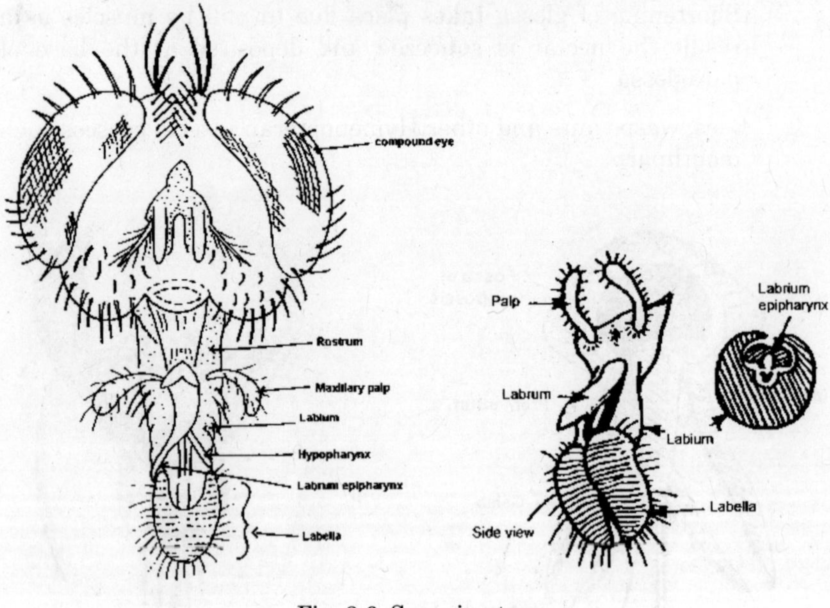

Fig. 8.6. Sponging type

Sponging (Cutting type)

1. Horse flies exhibits cutting sponging type.
2. Mandibles are shaped into sharp blades.
3. Maxillae are produced into long probing styles.
4. End of lower lip is modified into spong like structures.
5. Epipharynx and Hypopharynx collectively make a tube.
6. Maxillae and mandibles cut and tear the skin of mammals. Blood thus ooze-out is sucked by the sponge like end of lower lip. It is then conveyed to hypopharynx from where it passes to the oesophagus through the channel formed by epipharynx and hypopharynx.

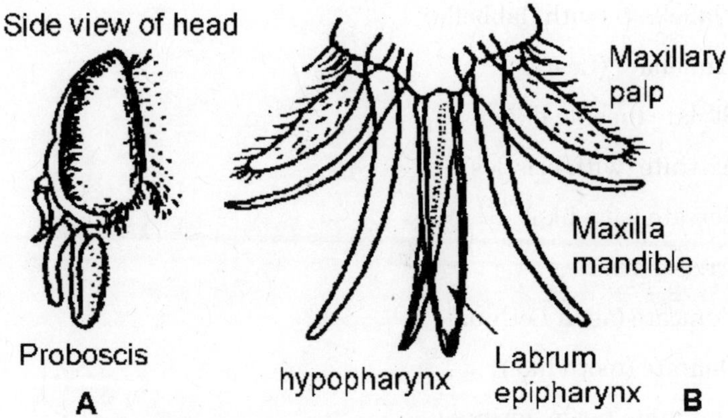

Fig. 8.7. Tabanus (sponging cutting type)

Permanent slide of different types of antennae (Fig. 8.8)

Remove the antennae from different insects, dehydrate and mount.

Atennae are sensory structures. They are situated between the compound eyes. Scape, pedicel and flagellum are the basic parts of the antennae.

Types

Filiform (thread like)

Setaceous (tapering)

Capitate (with head)

Clavate (clubbed)

Mucronate (tabanus like)

Plumose (hairy)

Moniliform (beaded)

Bipectinate (Comb like)

Tripectinate (Comb like)

Pectinate (Comb like)

Geniculate (elbowed)

Flabellate (with flabbella)

Lamellate (Leafy)

Stylate (needle like)

Aristate (with bristles)

Serrate (saw like)

Irregular

Tunicate (as in Lethius)

Dentate (tooth like)

Laminate (as in gyrinus)

Filiform

1. Segments are elongated
2. They are attached end to end.
3. They appears like thread.

Example : Cockroach

Setaceous

1. Each successive segments diminishes in diameter.
2. The distal part is tapering.

Example : Dragon fly.

Capitate

1. Segments from base to distal end are almost of equal size.
2. The last few segments are larger in size.
3. Commonly called head like.

Example : Tribolium and other beetles.

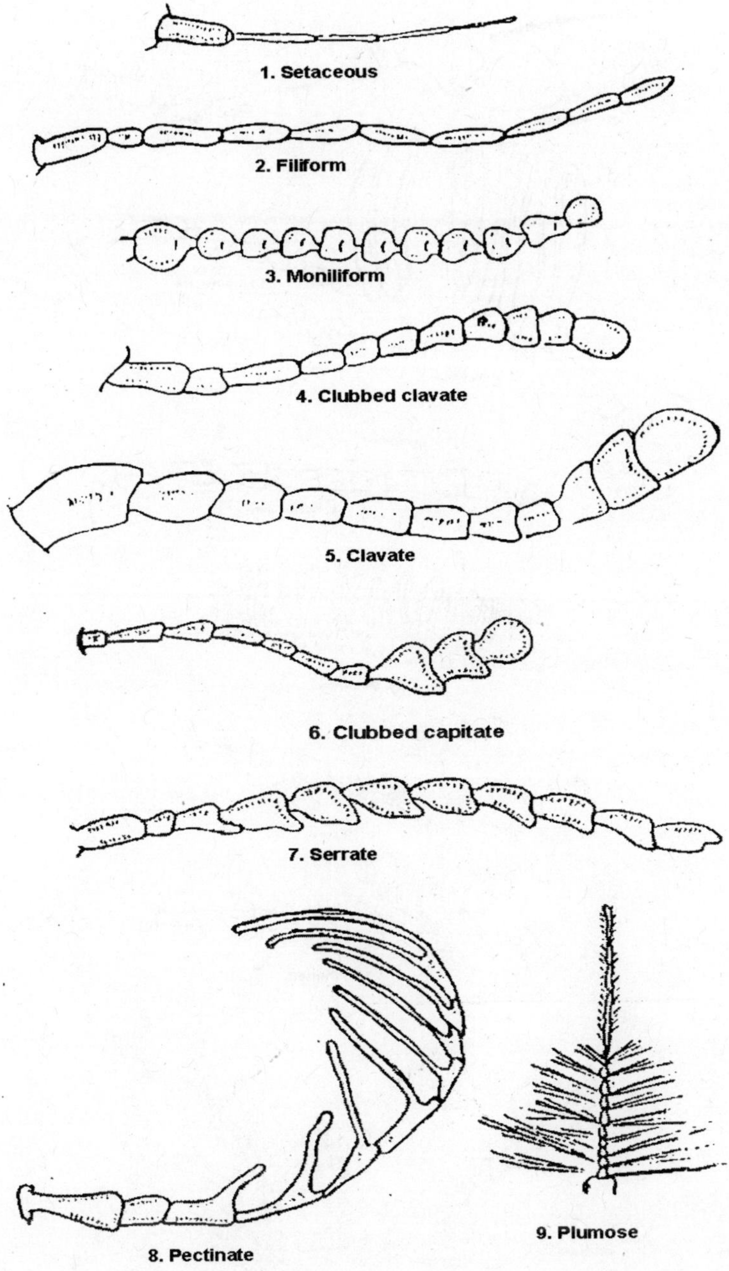

Fig. 8.8. Types of antenna

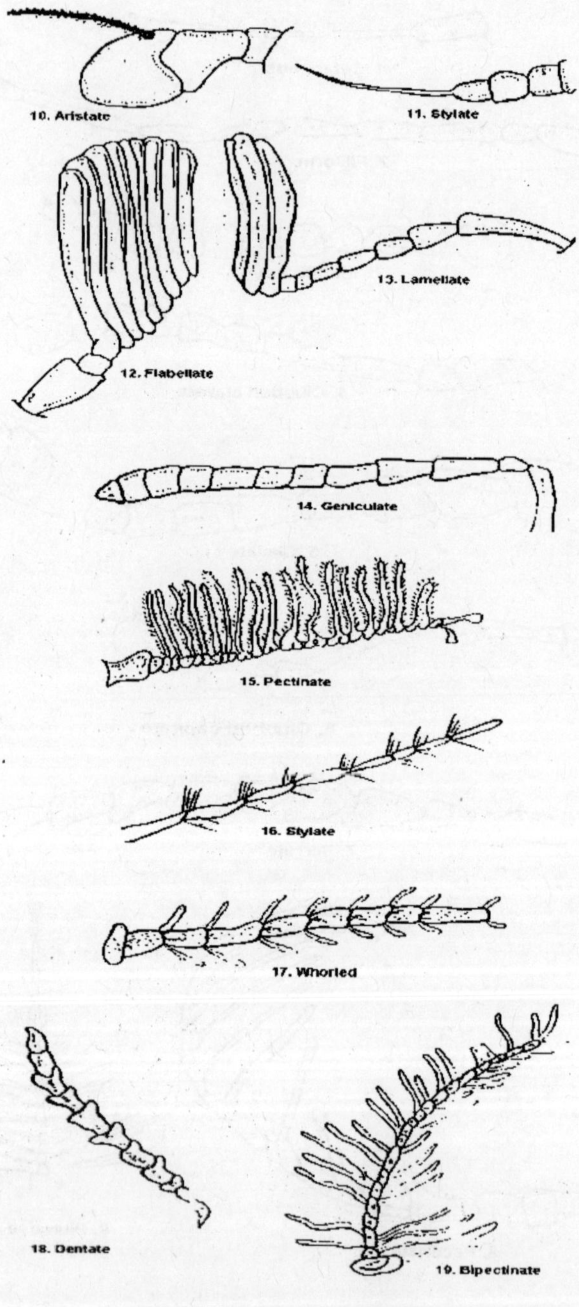

Fig. 8.8. Types of antennae.

Clavate

1. Segments gradually enlarges towards the distal end.
2. Appears club like.
3. Last few segments are round cone like.

   Example : Butterfly, some coleoptera.

Plumose

1. A whorl of hairs arises from each node.
2. It appears hairy.

   Example: Mosquitoes (male)

Pilose

1. Flagellum looks like a shaft flexible hair.
2. All segments alike fine tubular and without process.

   Example: Female mosquitoes.

Moniliform

1. It appears a string of beads.
2. Whole antennae is beaded.

   Example: Termites (Calotermes)

Pectinate

1. The side of the segments is produced into teeth of a comb.
2. Commonly called comb like.

   Example: Moths and male Callosobruchus.

Bipectinate

1. Comb is present on both side of the segments.

   Example: Australian moth

Tripectinate

Set of comb on three sides are present.

Example: Australian moth.

Geniculate

1. The body of the antennae makes a sharp curve or angle with scape.

2. It appears elbow shaped.

Example: Ant, Chalcid wasp and other Hymenoptera.

Flabellate

It bears a special organ flabella.

Example - Rhipiphoridae (coleoptera), Bombycoidea (Lepidoptera), Tenthredinidae (Hymenoptera).

Lamellate

1. Appears leaf like.

2. The terminal segment is composed of few lamellae like a leaves of a book.

Example: Dung beetle, Melolontha and beetles of Scarabaeidae.

Stylate

1. It appears needle like.

2. The last segment of antennae bears a bristle called style.

Example: Tabanus fly.

Aristate

1. The antennae bears large bristle called arista.

Example: Housefly.

Serrate

1. It appears saw like.

Example: Elaterid beetles, Buprestidae female callosobruchus.

Biserrate

When the saw like teeth are present on both sides.

Uncinate

1. The distal segment is sharp and curved.

Example: Odynerus.

Dentate

1. One side of the segments produced into teeth like projections.

Example: Stenochrorus.

Irregular

1. Pedicel is large and segments in flagellum are comparatively small.

2. Other segments toward distal end are smaller and the terminal segments is club shaped.

Example: Crocema.

Tunicate

1. The distal segment is belt like.

Example: Lethius.

Laminate

The segments are spirally arranged.

Example: Gyrinus.

Permanent slide of legs

Take younger stages of the insects and remove their legs, bleach dehydrate and mount.

Adaptive radiations in insects are witnessed by wide range of modifications in the legs. According to different mode of life they are adapted for walking, running, jumping grasping, swimming and clinging.

1. Ambulatorial legs

1. Adapted for walking.
2. Femur flat, tibia long.
3. Pretarsus is provided with claw and flattened arolium.
4. It can move perfectly on smooth and rough surfaces.

Example: Bugs, Dermaptera and some Beetles.

2. Cursorial Legs

1. Adapted for running.
2. Femur flat, coxa highly muscular.
3. Tibia and tarsi slender and elongated.

Example: Cockroach, Tiger beetle.

3. Saltatorial legs

1. Adapted for jumping.
2. Femur greatly enlarged
3. Trochanter is most modified

4. Raptorial legs

1. Adapted for catching the prey.
2. Elongated and spiny
3. Coxa long
4. Tibia can be accommodated into groove of femur and tarsus can be inserted into the apical hook of the femur.
5. Tibia apposable.
6. Reduced preapical tarsus.

Example: Preying mantis (forelegs), hind legs of some parasitic Hymenoptera fore legs of some aquatic beetles.

5. Natatorial legs

1. Adapted for swimming

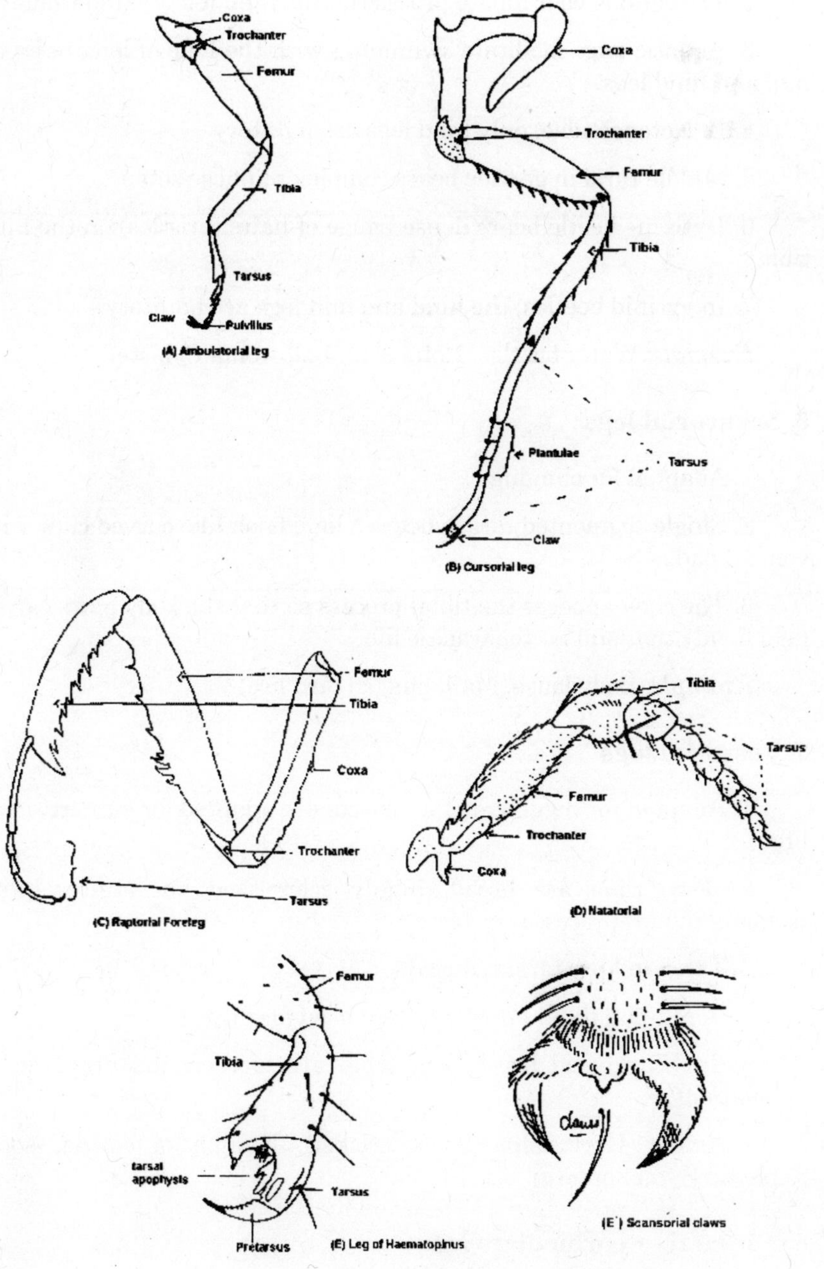

Fig. 8.10. (A-E[1]). Types of legs

2. The coreids bear foliate process on the hind femora and tibiae

3. Aquatic bugs facilitate swimming with the help of long hairs on mid and hind legs.

4. In Notonectidae, only hind legs are natatory

5. Middle tarsi in gerrids bears complex plumose setae.

6. Dytiscus beetle bears dense fringe of hairs (bristles) on the hind tibia.

7. In gyrinid beetles, the hind and mid legs are natatory.

Example: Water beetles, Notonecta, Belostoma Cybister.

6. Scansorial legs

1. Adapted for clinging

2. Single segmented tarsus bears a long hook like curved claw with ventral pad.

3. The claw apposes the tibial process so that the body hairs can be held firmly that suits ectoparasitic life.

Example: Body louse, Mallophaga (Bird lice)

7. Fossorial legs

1. Adapted for digging as the insects are adapted for subterranean life.

2. Fore tibia are broad, highly sclerotized but stumpy with terminal digital processes or teeth.

3. Tarsi and pre tarsi reduced.

4. Fore tibial digital processes are used for digging.

5. In Scarabaeid beetles the tibia are toothed and are used in excavations.

Example: Gryllotalpa (mole cricket), Nymph of cicada, some parasitic Hymenoptera.

8. Foragial or corbiculate legs

1. Adapted for assembling and collecting food.

2. Such legs are present in Honey bee.
3. In Honey bee the fore tibia bears a terminal spur known as fibula, the calcar which can be accommodated into proximal cavity of the tarsi.
4. Antennae are inserted into this cavity so that the pollens attached on antennae can be removed by the hairs (bristles) in this cavity.
5. Pollen basket or corbicula formed by the combs of metatarsal setae of both the hind legs. Comb is also called scopa.
6. In Dragon flies the legs are directed forward. They can cover mouth parts and pre oral cavity from the ventral side like a spoon.
7. Femur, tibia and tarsi bears spines and bristles. These long spines of one side combines to form a net for trapping collecting and transporting the prey like flies and mosquitoes.

Example: Honey bee and Dragon flies.

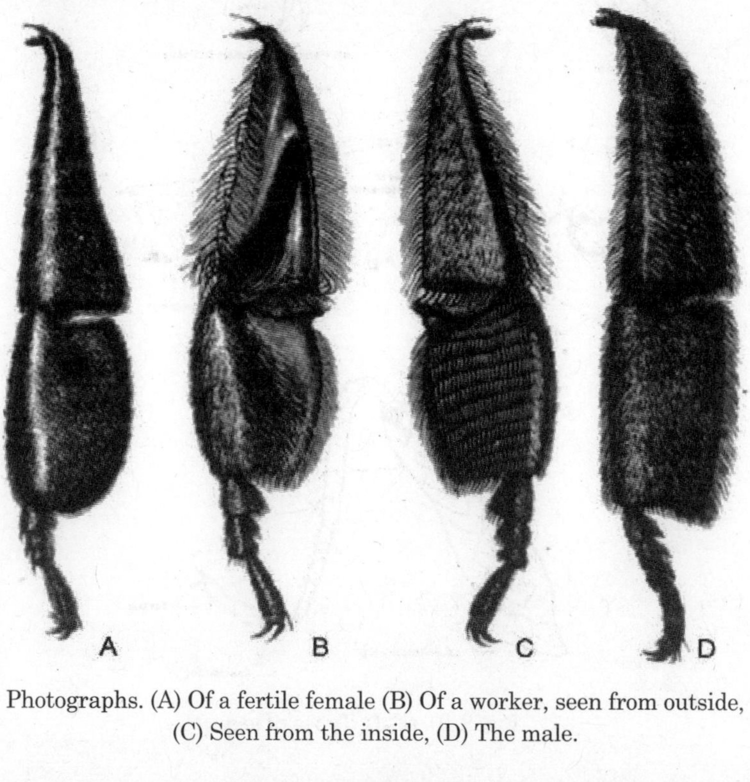

Photographs. (A) Of a fertile female (B) Of a worker, seen from outside, (C) Seen from the inside, (D) The male.

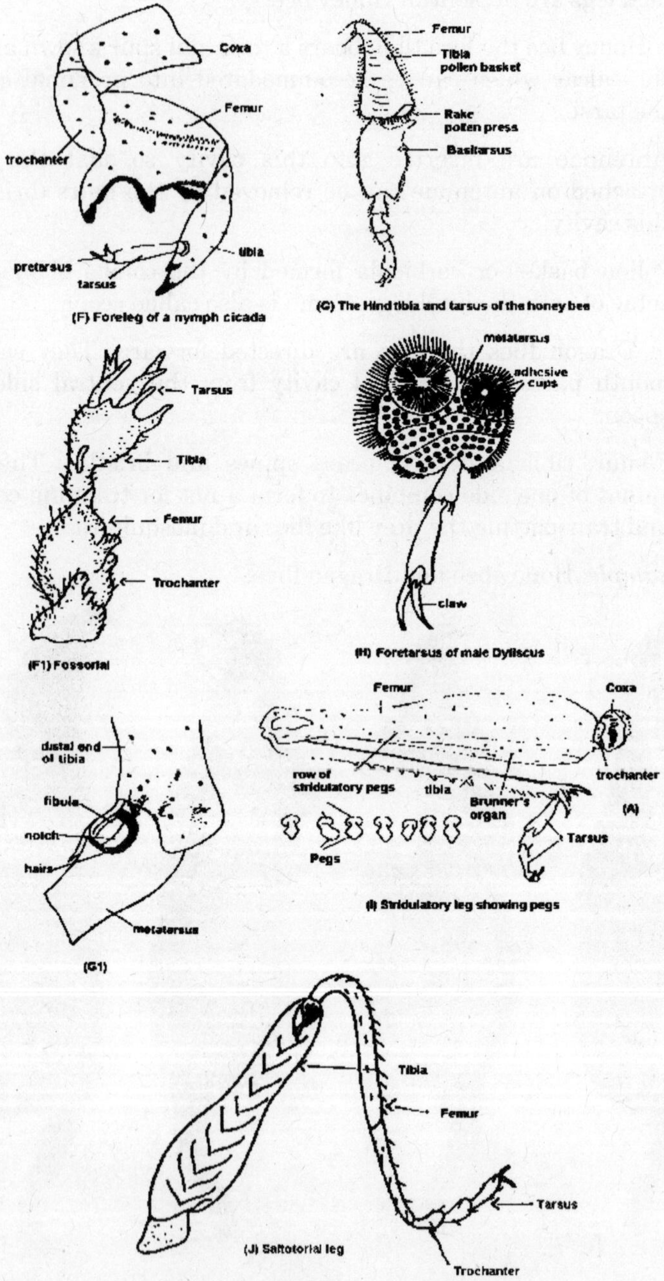

Fig. 8.10. (F-J). Types of legs.

9. Suctorial legs

The fore legs of some aquatic beetles are provided with suckers on the metatarsi and stumpy tarsomeres in males for holding the female during mating. A viscous secretion from the sucker in the tarsus is discharged so that they adhere to surfaces they are applied.

Example: Male of family Dytiscidae Hydrophilidae, Carabeidae, Cicindellidae.

10. Stridulatory legs

1. Adapted for sound production
2. A row of pegs situated on the inner side of each hind femur which would work against the outer surface of each tegmen (acrididae).
3. Special ridges on 2nd and 3rd abdominal segments which are rubbed by posterior femora (gryllidae).

Permanent slide of different types of wings

Remove the wings with the help of forceps, dehydrate and mount in D.P.X. /keep a record.

The basic plan of wing and wing venation is being described here.

WING AND WING VENATION

Insect possess two pair of wings. The front wings becomes specialized for the protection of hind wings when at rest and folded. In orthoptera e.g. grasshoppers the forewings are thickened parchment like and are called *tegmen*. Tegmen covers the hind wings in a roof like manner. In coleoptera, the forewings are sclerotized and leathery. Veins are absent. They are called *elytra*. Elytra of both sides meet along a straight line when folded over the body. In Hemiptera the fore wings are called *hemelytra*. The base is parchment like and the apex is membranous. In Diptera e.g. Houseflies the hind wings are modified into *halters*. They consist of a stalk and a knob. They are balancers.

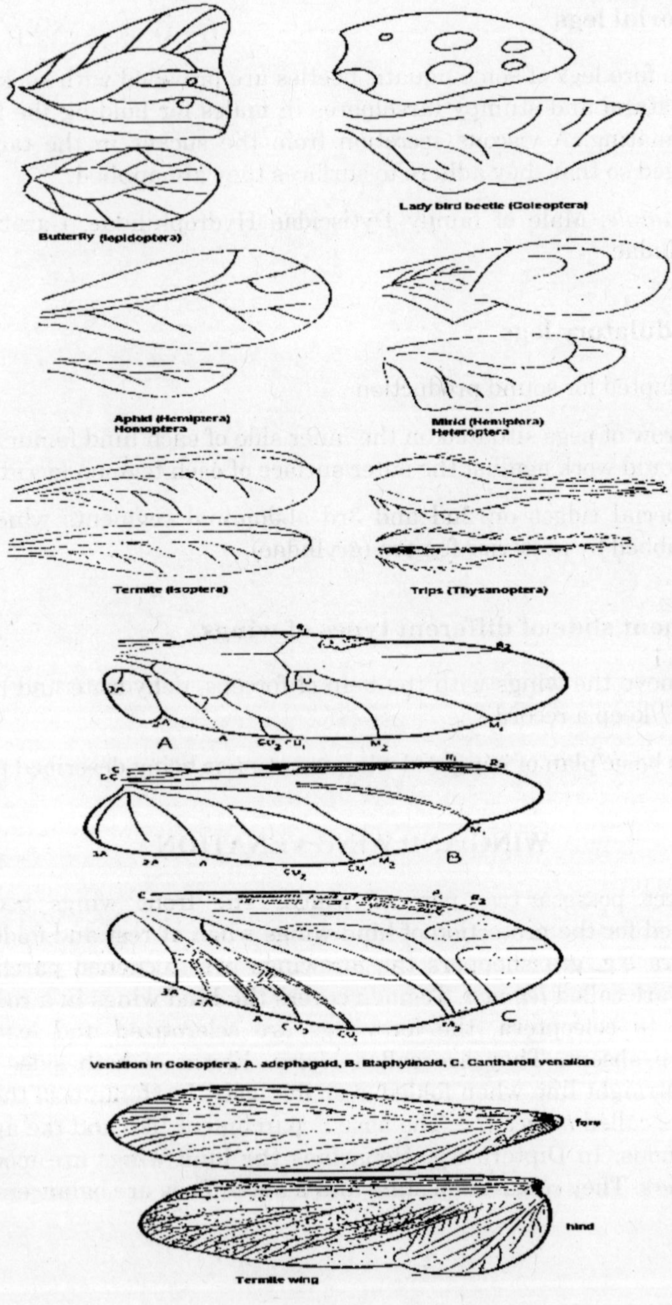
Butterfly (lepidoptera)
Lady bird beetle (Coleoptera)
Aphid (Hemiptera)
Homoptera
Mirid (Hemiptera)
Heteroptera
Termite (Isoptera)
Trips (Thysanoptera)
A
B
C
Venation in Coleoptera. A. adephagoid, B. Btaphylinoid, C. Cantharoil venation
fore
hind
Termite wing

Fig. 8.11.

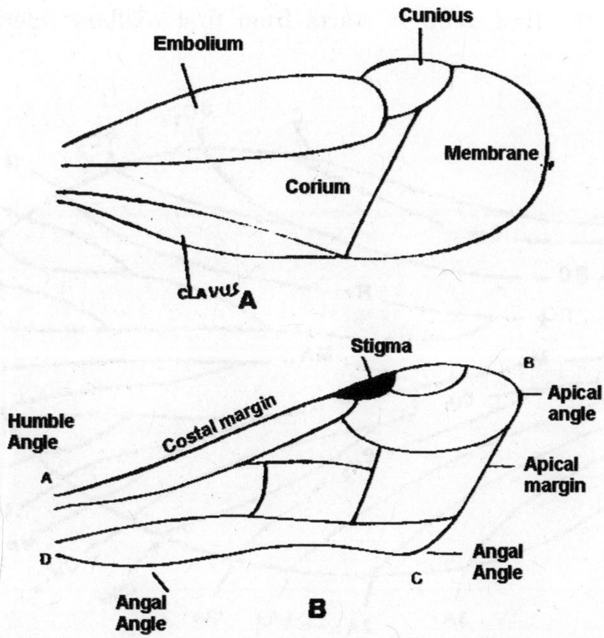

Fig. 8.11.A. Forewing of a Hemiptera, B. Wing of a Hymenoptera.

The arrangement of veins in a wing is called venation. Different orders have different venation. It is of great taxonomic importance Comstock and Needham (1898, 1918) constructed a hypothetical type of wing venation (Fig. 8.12). It was based on the study of numerous tracheae. It is presumed that all other types of wing venation have been derived from the hypothetical type. Generalized insects or most ancient insects representing simple system of venation, the *archetype* venation and it can be described as follows:

Costa (C)

It is the first anterior marginal vein hence the area of wing is called costal margin. It starts from the wing base (humeral plate). It is unbranched and convex. It sometimes bear stigma or pterostigma on fore and hind wings of odonata and only on forewings of Hymenoptera, Psocoptera and Mecoptera.

Sub-costa (Sc)

It is the second vein and is divided distally into two branches, the outer and inner designated as Sc_1, and Sc_2. In absence of costa it represents the first vein. It starts from first axillary sclerite. It is concave.

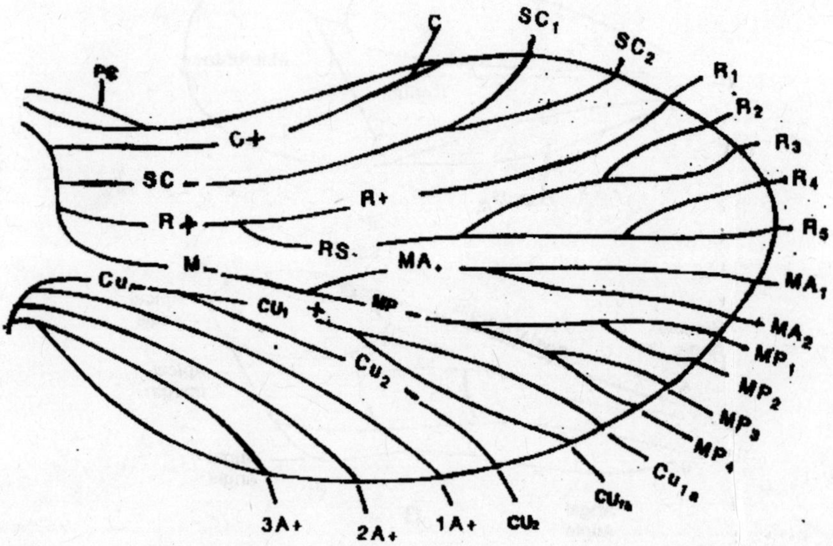

Fig. 8.12. Hypothetical wing venation, 1A+2A+3A + branches of anal veins, MA - anterior media, MP - posterior media, MA1, MA2, MP1, MP2, MP3, MP4 - branches of the anterior cubitus, CU1b - peosterior cubitus, PC - pre costa, C - costa, SC - sub costa, R + radials, M = media, CU - cubitus, SC1 - subcosta 1, SC2 - sub costa 2, R1, R2, R3, R4, R5 - branches of radius, CU1, CU2, branches of cubitus, R - radius.

Radius (R)

It is the third vein which starts from the 2nd axillary sclerite. Before its division into two branches it is convex. The first branch, the outer branch (R_1) runs directly towards outer margin. The second branch is called radial sector (R_s) which is concave and gives branches R_2, R_3, R_4 and R_5.

Media (M)

It starts from medial plate and is fourth vein. It divides into two branches viz. anterior media (MA) which is convex and further branched into MA_1 and MA_2 and posterior media (4-branched, MP_1, MP_2, MP_3 and MP_4 and are concave). Anterior media is present in Odonata.

Cubitus (Cu)

It starts from the distal medial plate and is fifth vein. It divides into convex anterior (Cv_1) and concave posterior (Cv_2) branches. The anterior cubitus is two branched, the anterior (Cu_{1a}) and posterior (Cu_{1b}) and posterior is unbranched.

Anal vein (A)

It starts from third axillary sclerite which constitutes a group of veins. The anterior most anal vein is also called post cubitus (pC_u). Their number is fluctuating in different orders. They are generally unbranched convex.

Jugal vein (J)

Though their number varies, but first and second jugal veins are commonly found.

Besides these longitudinal veins, a number of cross veins are also found between them. The principal cross veins are Humeral (h) run between sub-costa and costa; the Radial (r) extending from R_1 to R_5 (Radial sector); the Radio, medial (rm) between Radius (R) to Media (M); the Medial (m) between MP_2 and MP_3 and the Medio-cubital (m-cu) from MP to Cu.

WING COUPLING APPARATUS

Primitive type

The jugal lobe at the base of forewings and a humeral lobe at the base of costal margin of hind wing which contain setae, form a simple primitive type of coupling apparatus.

Example: Mecoptera (Fig. 8.13)

Jugate and Frenate

It is found in Neuroptera, Trichoptera, Lepidoptera. Jugal lobe is long on the fore wing or developed frenulum and retinaculum like bristles.

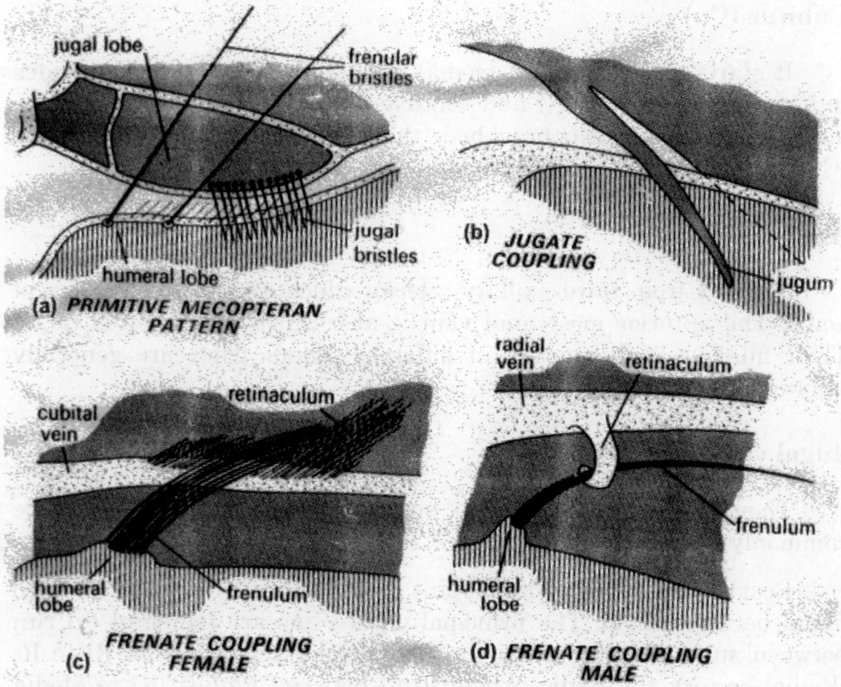

Fig. 8.13. Wing coupling mechanisms.

Hamulate

It is found in Hymenoptera. A row of hamuli (small hooks) on middle of costal margin of the hind wing and a thick cuticular fold along hind margin of the forewing developed.

Permanent slide of different types of eggs

Permanent slide of different eggs can also be prepared. Use a cavity slide, dehydrate and mount in D.P.X.

Brief description of some eggs is given below.

1. Eggs of chrysopa

They are laid on leaves but raised on stalks which may be upto 15 mm high. The stalk is produced as a viscous material which hardens in the air.

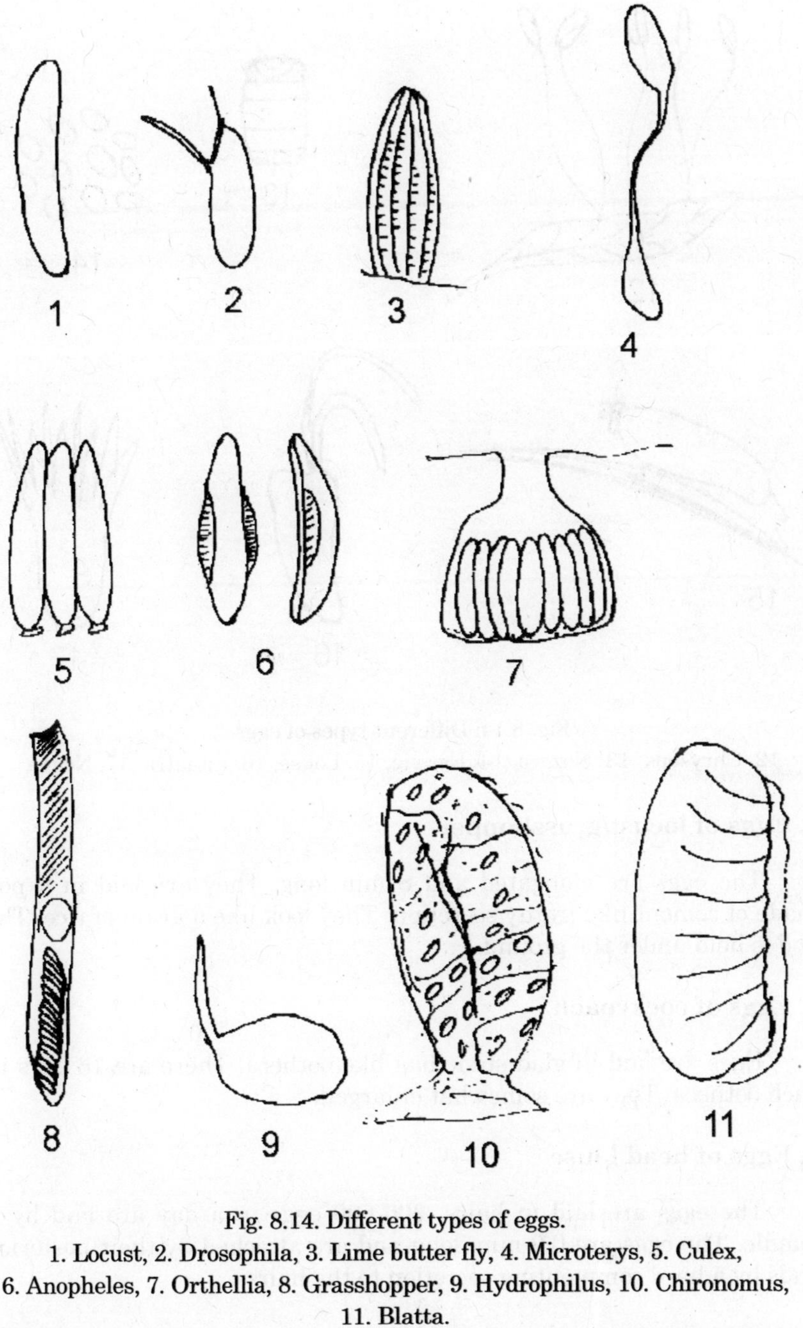

Fig. 8.14. Different types of eggs.
1. Locust, 2. Drosophila, 3. Lime butter fly, 4. Microterys, 5. Culex, 6. Anopheles, 7. Orthellia, 8. Grasshopper, 9. Hydrophilus, 10. Chironomus, 11. Blatta.

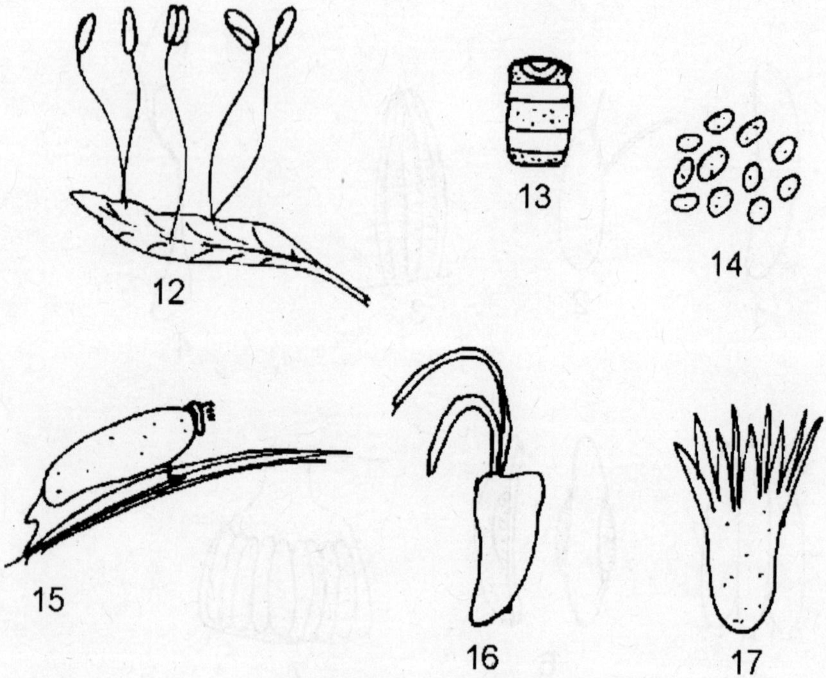

Fig. 8.14. Different types of eggs.
12. Chrysopa, 13. Nezara, 14. Earwig, 15. Louse, 16. Ranatra, 17. Neepa.

2. Eggs of locust/grasshopper

The eggs are elongated and 8 mm long. They are laid in a pod made of cement like frothy secretion. They look like a grain of rice. The pod is held under the ground.

3. Eggs of cockroach

Eggs are laid in glad stone bag like ootheca. There are 16 eggs in each ootheca. They are somewhat enlarged.

4. Eggs of head louse

The eggs are laid in hairs. 200-500 eggs in a day are laid by a female. The eggs are 0.9 mm. long and are attached by their posterior ends by a hard commenting secretion to the hairs.

5. Eggs of Anopheles and Culex

Culex constructs a raft of 150-300 eggs between the hind tarsi which lie flat on the surface. The eggs float in an upright position. Anopheles lay eggs singly on the surface and the position in which the eggs float. Air filled floats in the chorion is present.

6. Eggs of Musca and Drosophila

Eggs are laid in food. The eggs are creamish white and a millimeter long.

7. Eggs of Nepa and Ranatra

Nepa and Ranatra lay eggs in floating vegetation. Respiratory horns remains out of water.

8. Eggs of Pieris

The eggs are spherical and sculptured.

9. Eggs of Earwigs

Female lay a cluster of eggs in under ground chamber and mount guard over them. Eggs are regularly picked up and cleaned by female to remove any fungal spores. The eggs are spherical and white in colour.

Permanent slide of different types of larvae

Slide of different larvae can be prepared. First instar larvae are best suited for permanent slides. Dehydrate and mount in DPX. Given below is the brief description of different types of larvae.

1. Eruciform

Thoracic legs well developed or vestigial sluggish and robust. Found in some Coleoptera, Hymenoptera.

2. Apodous

Legs absent. Head capsule is reduced. Lives in abundant food.

Example: Diptera, Hymenoptera and Coleoptera (weevils).

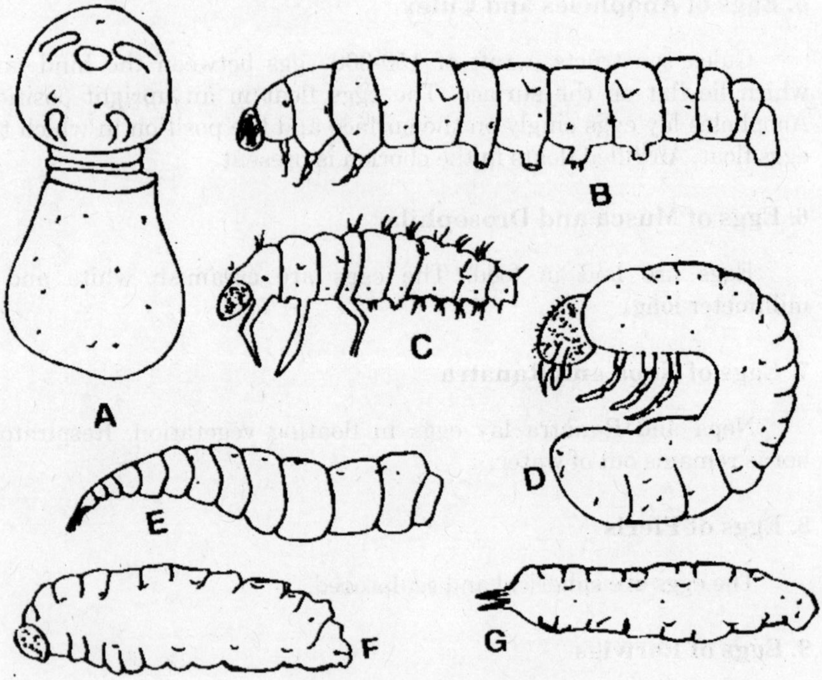

Fig. 8.15. Different types of larvae.
A. Protopd, B. Polypod, C. Campodeiform, D. Scarabeifom, E. Acephalous, F. Eucephalous, G. Hemicephalous.

3. Scarabeiform

Head well developed, C shaped. Prolegs absent but thorocic legs present. Lives on dung or decayed vegetation.

Example : Dung beetle.

4. Cylindrical

Worm like elongated, no appendages.

Example : Wood boring beetles, Honey bee and other Hymenoptera.

5. Campodeiform

Legs long and body flattened. Mandibles short sickle like, antennae and cerci well developed.

Example: May fly, Odonata, Plecoptera, Neuroptera & Trichoptera.

Types of Pupae

It (pupa) occurs in all holometabolous insects. It is a resting inactive, non feeding preimaginal stage of development. Classification as given by Hinton, 1946 is given below.

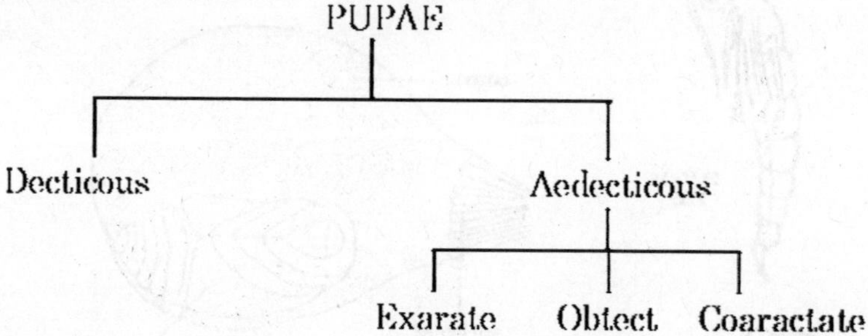

Decticous (articulated mandibles, mobile)

They possess powerful mandibles that are scleortized and are innervated by pharate adult muscles. They are used to escape from the cocoon. Decticous is a primitive type of pupa and is always exarate i.e. appdendages are not adhered to the body and can be used for locomotion e.g. Trichoptera, Mecoptera, Neuroptera and some families of Lepidoptera (Micropterygidae and Eriocraniidae).

Adectious (mandibles immobile)

They have much reduced mandibles and are useless for escaping from the cocoon.

Adecticous exarate

Appendages are free (the wings rudiments legs and antenna are free) e.g. Strepsiptera, Siphonaptera most of coleoptera, Hymenoptera Cyclorhapha and Bracheycera of Diptera.

Adecticous obtect

The appendages are fastened to the body of pupa along its length e.g. Higher Lepidoptera. Some Coleoptera, Nematocera of Diptera. Chalcidoidea of Hymenoptera.

Adecticous coarclate

The last larval moult is retained and becomes hardened to form a protective covering for pupa that developed inside. It is called *puparium* and is larval shaped and segmented. There is no external sign of legs, antennae wings. It is characteristic of cyclorrhaphous Diptera.

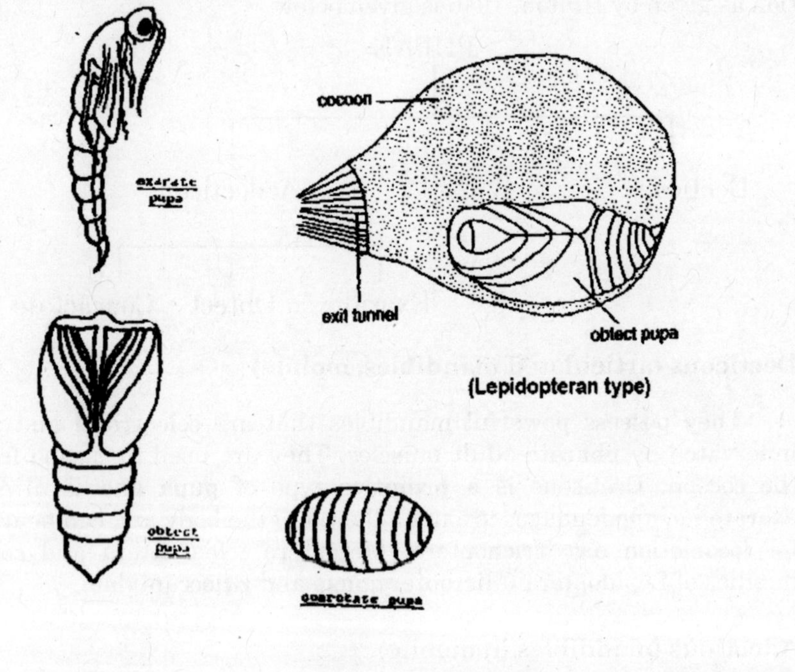

Fig. 8.16. Different types of pupae.

II. DISSECTIONS

1. MINOR DISSECTIONS

(i) Tympanum

It is situated at the base of fore tibia in Grylloidea and Tettigonoidea. The tympanic membranes are protected by forwardly projecting folds of the tibial cuticle. The whole cavity of the tympanic membranes is occupied by a branched trachea. Tympanal organs are also present in Locust, Cicada and certain moths (either side of first abdominal segment in locust; posterior end of first abdominal segment in cicada; posterior part of metathorax in noctuid moths).

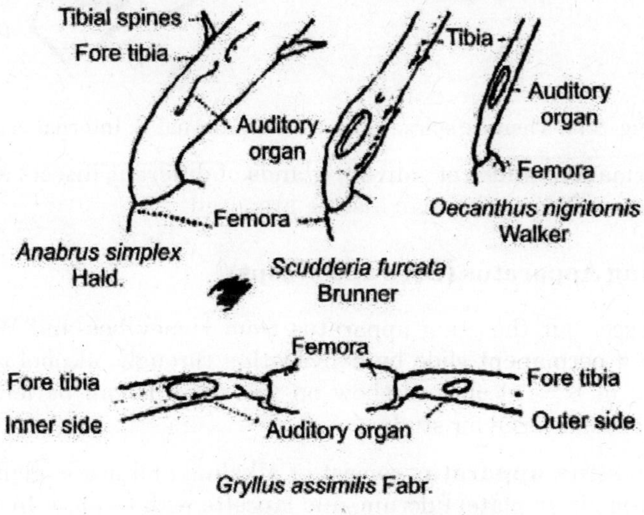

Fig. 8.16(a). Tympanum on fore tibia

(ii) Spiracle

Spiracles are the external opening of the tracheal system. They are lateral in position on the pleura.

Locate and take out the spriacle from between meso and metathorax and make a permanent slide in DPX. Spiracle in grasshoppers is closed by two movable semicircular valves which are unsclerotized except at the hinge and are thickened basally to form a pad into which the muscle is inserted. The pulling of these muscles facilitates its rotation and closing.

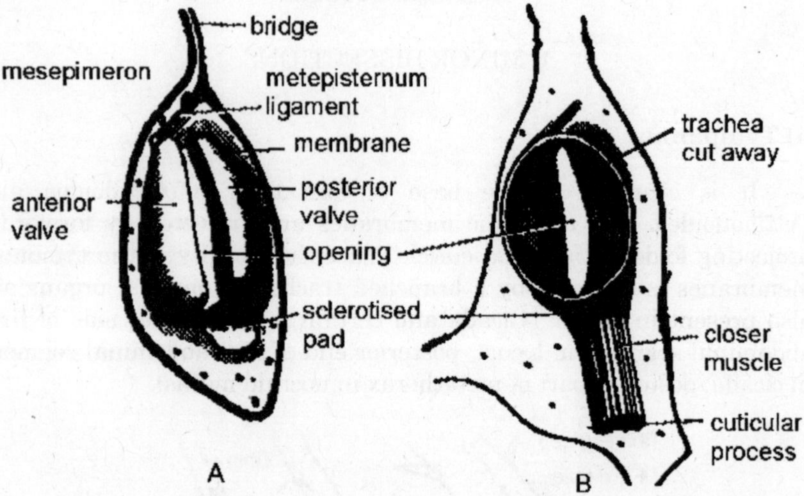

Fig. 8.17. Thoracic spiracle (Locust) A. External B. Internal view.

Permanent slides of salivary glands of different insects and sting apparatus of Honey bee can also be prepared.

(iii) Sting Apparatus (Bees and Wasps)

Dissect out the sting apparatus from Honey bee and Wasp and prepare a permanent slide by dehydrating through alcohol grades or preserve in 80% alcohol or show on the slide. It can be left in 70% alcohol as a material for study.

The **sting apparatus** consist of alkaline and acidic glands, IXth tergite (quadrate plate) fulcrum, and lancette with terebra. In aculeata, the ovipositor is no more concerned with depositing the eggs. The two accessory glands lie one behind other and are modified as poison glands behind the genital opening. The stylet is modified into sting and lies in sting chamber. The secondary function of the ovipositor is boring, piercing, sawing, stinging and paralysing the prey. It is absent in drones.

In Stinging, the muscles of IX tergites drive the style and stylet sheath into the victim, the secretion of the poison glands pass down to the wound.

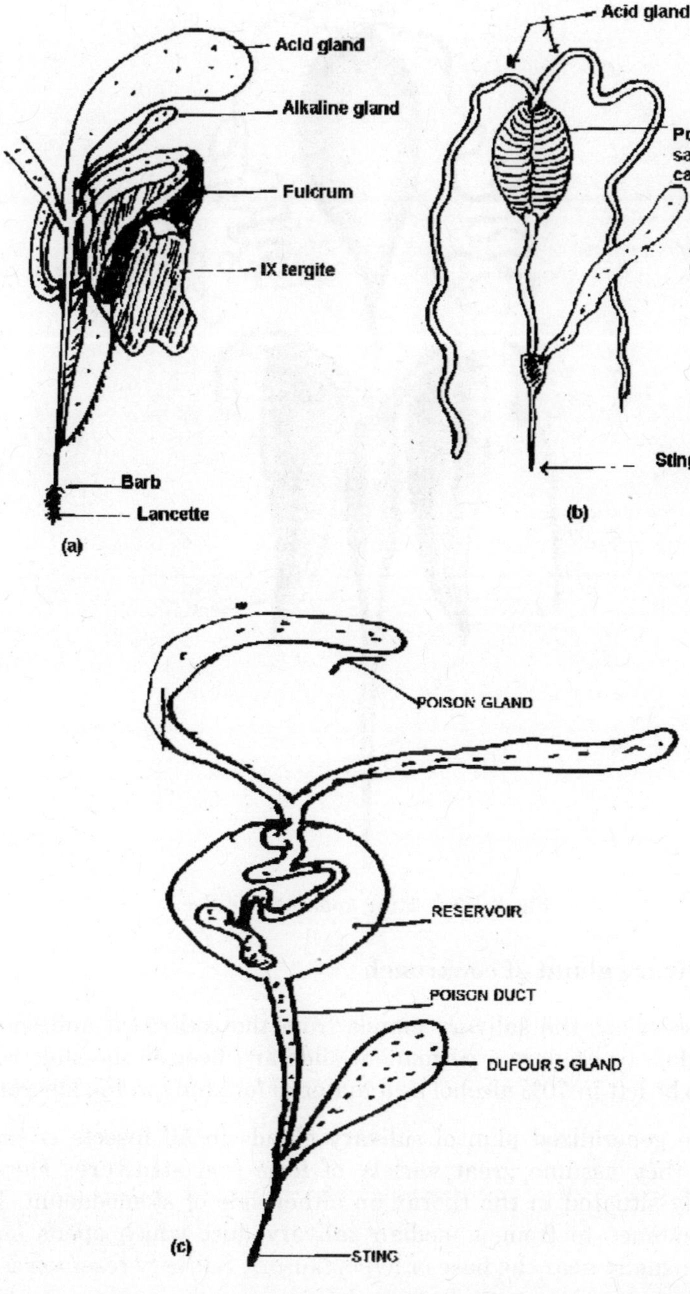

Fig. 8.18. (a) Sting apparatus of honey bee. (b) Wasp (c) Poison gland of ant.

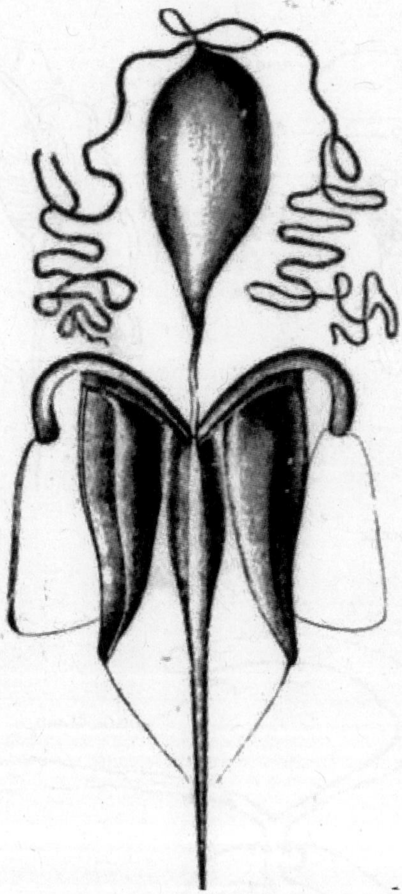

Fig. 8.18(a). Sting apparatus of Bee.

(iv) Salivary gland of cockroach

Dissect out the salivary glands from the cockroach and show in watch glass or prepare a permanent slide and keep in the slide box. It can also be left in 70% alcohol as a material for study in the laboratory.

The generalized plan of salivary glands in all insects is similar though they assume great variety of form and structure. They are generally situated in the thorax on either side of stomodaeum. Their duct combines to from a median salivary duct which opens on the labium usually near the base of hypopharynx. Salivary reservoir is also present.

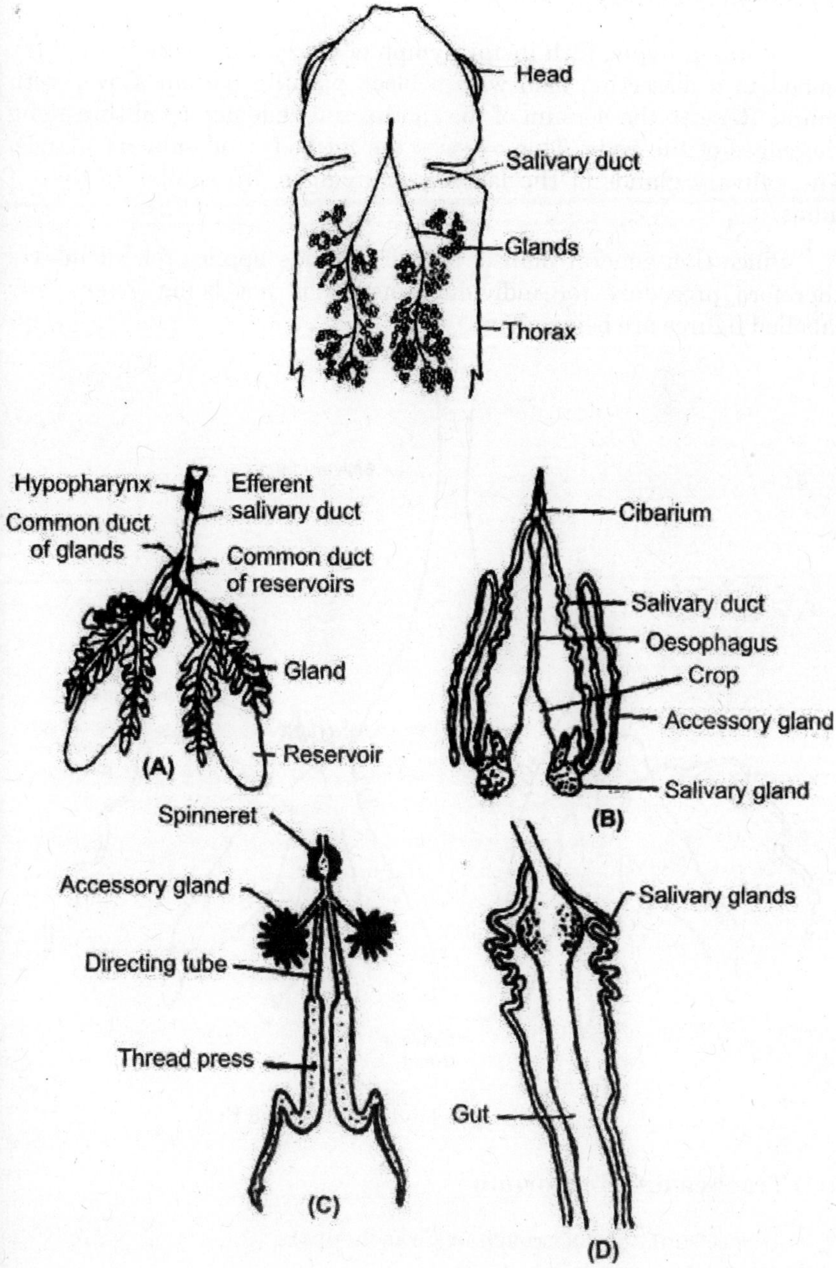

Fig. 8.19. Salivary glands A. Periplaneta, B. Dysedercus, C. Saturnid silk worm D. Musca, and E. Locust.

(v) Salivary glands of a bug

Obtain a living, fifth instar nymph of *Oncopeltus* (milk weed bug). Imbed in a dissecting dish with a black paraffin bottom. Cover with saline. Remove the dorsum of the thorax and abdomen by silting along the edges of the body. This exposes the fat body and salivary glands. The salivary glands of the last instar nymphs are similar to that of adult.

Since this general plan of salivary glands applies for all insects therefore procedure for individual insects is not being given, only labelled figures are being given.

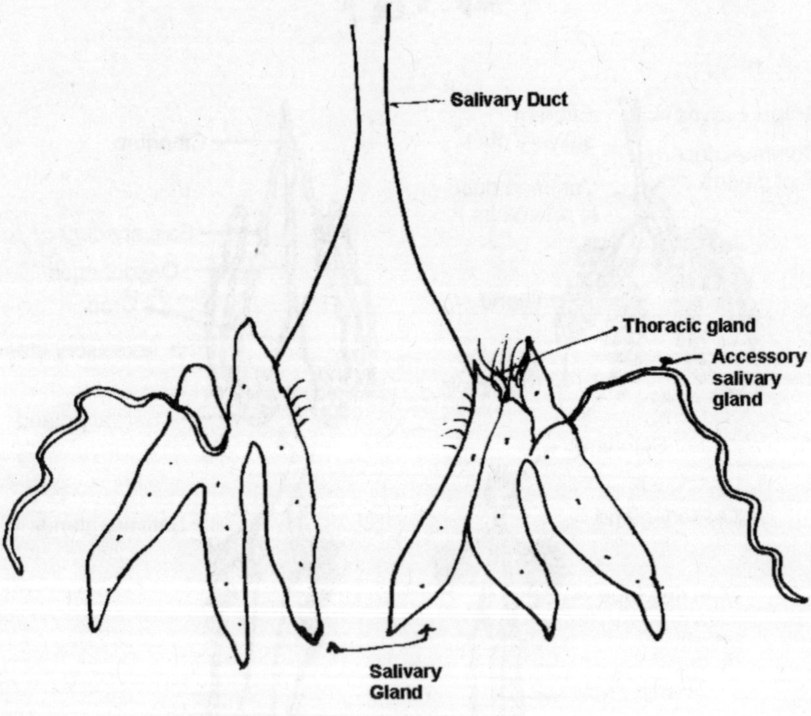

Fig. 8.20. Salivary gland of Milk Weed Bug.

(vi) Trachea and Tentorium

Dissect out the cockroach or Grasshopper.

Pickup single trachea and mount it. Dissect the tentorium as per the instructions given by the teacher.

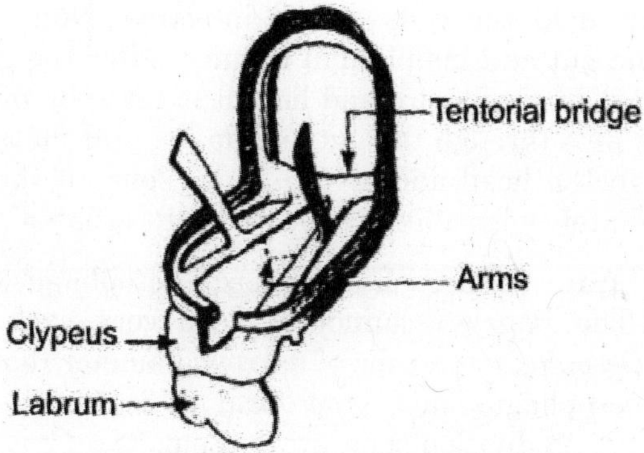

Fig. 8.21. Tentorium.

Tentorium

It provides rigidity to the head. It is a frame work of apodemes which serves to afford attachment for muscles. It consists of a central bridge (carpotentorium) from which three pair of arms extend to the cranial wall. The *anterior arms* are intagimated from the anterior tentorial pits on the postoccipital suture. The *dorsal arms* extend to the capsule between the antennae and the lateral ocelli. The *Posterior arms* cannot the carpotentorium with the rear of the capsule.

2. MAJOR DISSECTIONS

The success of the dissection depends on individual patience and fine and sharp dissecting instruments. Follow the instructions of your teacher. Use dissecting binoculars. Never cut away any part until you are quite certain what it is you are removing.

(i) The Alimentary canal and associated organs of the cockroach

Procedure : Obtain a living adult cockroach. Grasp it gently but firmly and remove the wings with small scissors. Lay down the insect on its ventral surface (exposing dorsal side towards you) and insert a pin through the head and fix obliquely. Pour some physiological saline. Gently tease off the dorsal plate of the abdomen and thorax. The rectum lies directly beneath the integument. The crop and salivary glands are very superficial and special care should be taken while removing the pronotum. Expose the gut completely and carefully clip the trachea that hold the coils of the intestine. Notice the strong movements of the gut and malpighian tubules. After the gut has been uncoiled, extend it in a wide arc and hold it is place by minuten pins. Do not place the pins through the gut and do not pull the gut free from its connections in the head and at the anus. Pour off the saline and replace it immediately with a fixative or with 70% ethanol.

The **fore gut** consist of pharynx, oesophagus, crop and proventriculus. The *Pharynx* cannot be seen very well in a dorsal dissection. The *Oesophgus* is a tube of narrow diameter that follows the pharynx. The oesophagus and crop form a continuous, gradually widening sac, with no division separating them. The *Crop* extend well into the abdomen. The *Proventriculus* is cup like and is applied to the blunt end of the crop. Gently push the finger like gastric caeca aside to get a better view of the proventiculus. The gastric caeca belongs to the mid gut.

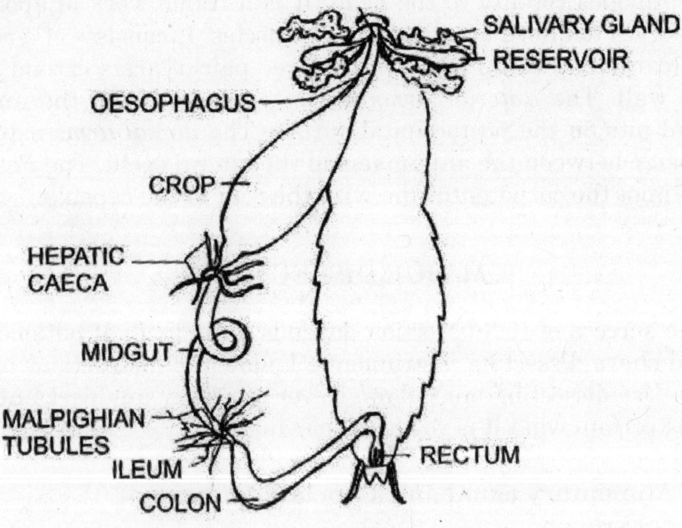

Fig. 8.21(a). Alimentary canal of cockroach (*in situ*)

The **mid gut** extends from proventriculus to the bases of malpighian tubules. There are usually eight gastric caeca, these are not of equal size, but all of them are attached to the anterior border of the mid gut. The *Malpighian tubules* are attached at the juncture of the mid gut and the hind gut.

The **hind gut** consist of the anterior intestine and rectum. Just behind the Malpighian tubules is a short-*pylorus*. The remainder of the anterior intestine is a longer tube which tapers slightly until it reaches the rectum. The rectum terminates at the anus.

(ii) The Alimentary canal of a Bug (Milk weed bug)

The milk weed bug *Oncopeltus fasciatus* (Dallas) serves to illustrate some of the specializations of alimentary canal of Heteroptera. These specializations are (i) simplification of the fore and hind gut, and (ii) regional specialization of the mid gut.

Procedure : Obtain a living milk weed bug. Remove the wings and imbed ventral side down in a paraffin dissecting disk. Cover with physiological saline. Remove the thin continuous dorsum of the abdomen. Expose the coiled gut. Now carefully remove the tergal plates of the thorax. The salivary glands and a large portion of the first division of the mid gut lie directly beneath the thoracic tergal plates. Be careful not to damage these parts. The digestive tract is about one and half times the length of the body, it is quite straight except about one and half times the length of the body. It is quite straight except in the regions of the second and third division of the mid gut where it is coiled upon itself. Carefully pluck the trachea that hold these coils and arrange the uncoiled gut in the dissecting dish Pour off the saline and add a fixative or 70% alcohol.

The **foregut** comprises of *pharynx* and *oesophagus*. Pharynx is elongate and only very slightly enlarged. The *oesophagus* comprises the greater part of the foregut.

The **mid gut** is composed of four distinct divisions. The *first division* of ventriculus generally extends well into the abdomen. The *second division* composed of convoluted tubes of relatively small diameter. The *third division* is oval to spherical in shape when full and usually contains a brownish oily mass. The *fourth division* is narrow and short. Histological differences in the four regions justify this anatomical divisions of the mid gut and probably reflects important physiological specializations.

The posterior limit of mid gut and anterior limit of hind gut are not easy to establish because the pyloric valve is lacking. Four long malpighian tubules (2 on each side) empty independently into the pylorus. The rectum is bulbous and leads to the anus by way of sort rectal tube (Fig. 8.22a & b).

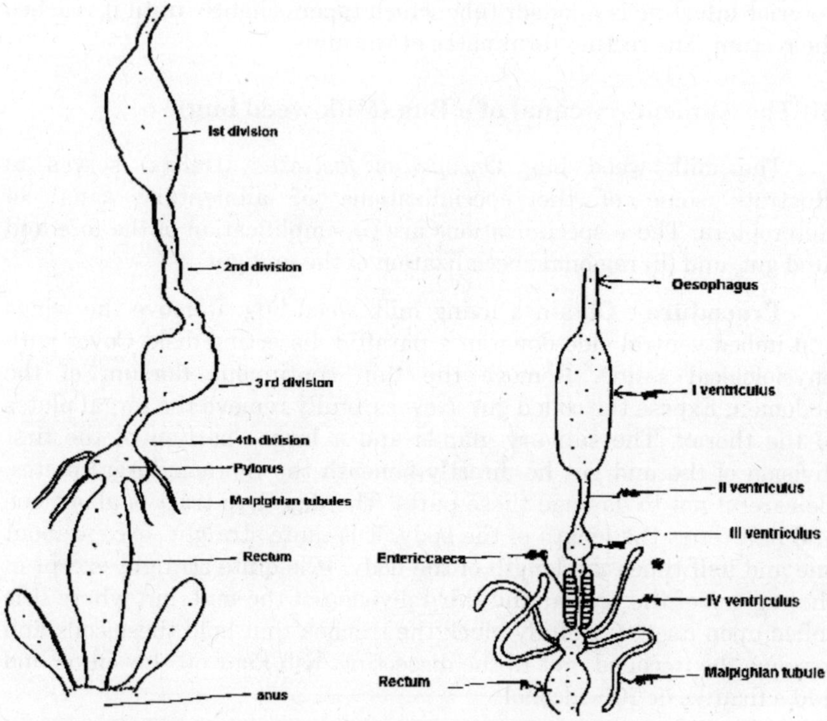

Fig. 8.22(a) Alimentary canal of Milk weed bug (Diagrammatic)
(b) Alimentary canal of Heteroptera (Bug) (Diagrammatic)

(iii) Alimentary canal of Homoptera (Aphid)

The narrow oesophagus extends from head to the *filter chamber*. Note the salivary glands at the sides of the oesophagus. The oesophagus leads to the **1st division of the ventriculus**. This part of the mid gut is an opaque white curved structure lying within the filter chamber. The **2nd division of the ventriculus** is bulbous and lies outside the filter chamber. The long coiled tube which extends from the 2nd division of the ventriculus back to the filter chamber is the 3rd division of the ventriculus. The outer wall of the filter chamber is probably formed by the posterior part of the mid gut and anterior end of the hind gut. It is difficult to determine the exact boundary between the mid and hind guts because the pyloric valve and malpighian tubules are lacking. The voluminous hind gut leaves the filter chamber and proceeds directly to the anus (Figs. 8.23a & b).

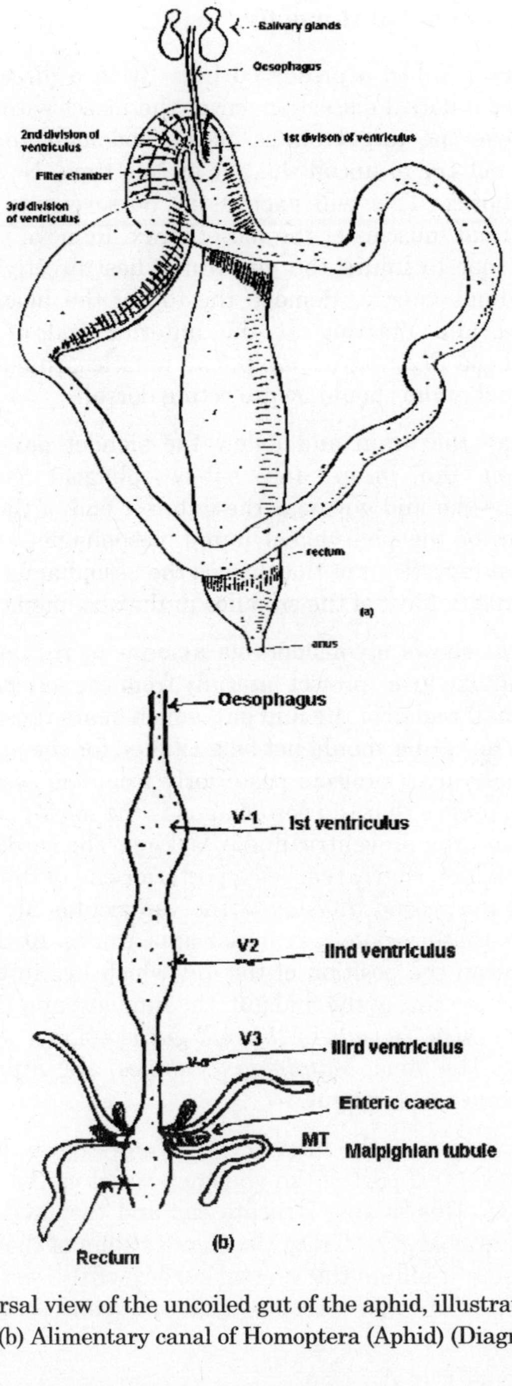

Fig. 8.23. Dorsal view of the uncoiled gut of the aphid, illustrating the filter chamber (b) Alimentary canal of Homoptera (Aphid) (Diagrammatic)

(iv) Alimentary canal of Horse fly

Procedure : Imbed a preserved horsefly in a dissecting dish in order to prepare a dorsal dissection, cover the insect with 70% alcohol. Carefully remove the tergal plates of the abdomen and expose the coiled gut. Do not try to uncoil the gut at this time. Now remove the terga of the thorax. This will expose the dorsoventral muscles and dorsal longitudinal muscles of the mosothorax. Remove these muscles but be careful not to injure the gut which lies directly beneath the dorsal longitudinal muscle. Remove the top of the head and dissect down to the brain. Starting at the anterior end of your dorsal dissection, find the *Pharyngeal pump*, two muscles, the dorsal dilators of the pharyngeal pump should be projecting dorsal.

Tease away the brain and follow the slender *oesophagus* as it passes between the brain and sub-oesophageal ganglion. The oesophagus joins the mid gut near the anterior end of the thorax. The foregut consists of the pharyngeal pump, oesophagus and crop. The *crop* is a ventral diverticulum that leaves the oesophagus just before it reaches the mid gut. Most of the crop lies in the abdomen.

The *midgut* shows a considerable amount of regional differentiation. Two wing like arms project laterally from the anterior end of the mid gut, the small region of the mid gut, which bears these processes is called *cardia*. The cardia should not be mistaken for the proventriculus. The oesophageal valves projects posteriorly from the oesophagus into the cardia, this proves that the oesophagus is the most posterior region of the foregut and the proventriculus is lacking. The cardia is joined to a straight tube which nearly reaches a posterior end of the thorax. Both the cardia and the *second division* of the ventriculus are composed of cells which give these regions a granular appearance. At this point it is necessary to uncoil the position of the gut which lies in the abdomen. This will expose the rest of the mid gut, the hind gut and the main body of the crop. The *third division of the mid gut* is smooth walled narrow diameter tube. The four *Malpighian tubules* are attached at the junction of the mid and hind gut.

Before passing on to the hind gut, carefully expose the transverse sac lying in the ventral part of the abdomen which is the main storage area of the crop. This is very delicate sac and the fat body must be picked up with utmost care. Trace the slender stem of the crop anterad to the point where it enters the ventral surface of the oesophagus. The *hind gut* consists of anterior intestine, the *rectum* and rectal tube.

The anterior end of the intestine forms a cup like *Pylorus*. The rectum is distinctive in possessing rectal papillae. The rectum tapers into a rectal tube which leads to the exterior by way of the anus (Fig. 8.24).

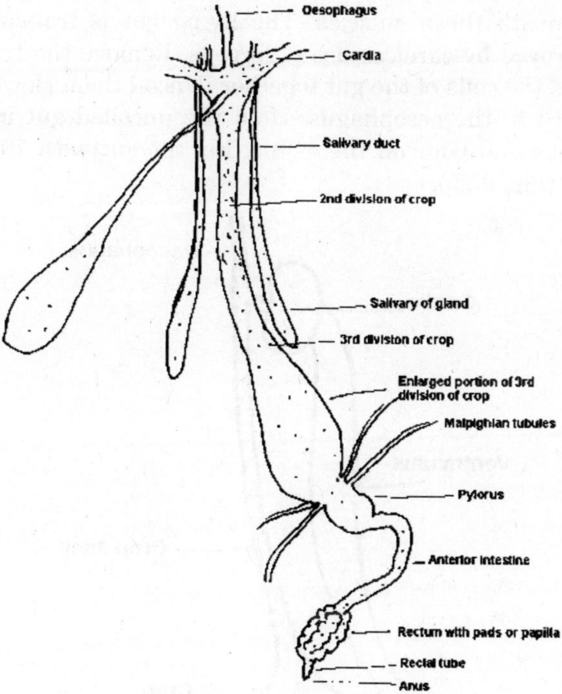

Fig. 8.24. Alimentary canal of Horse fly (Diagrammatic)

(v) Alimentary canal of Housefly

Note : This dissection will test your dissecting ability and interpretive powers. It is difficult one. The gut of the housefly is basically similar to that of horse fly therefore, no description is being given here. Only the procedure is dealt-here with.

Procedure : Anaesthetize a living housefly with carbon dioxide or inactivate them by keeping in the freezer for some time. Remove the wing and imbed the fly on its right side in a dissecting dish. Fix it in the paraffin. Use 30x magnification and with the help of sharp forceps carefully remove the side of the thorax and abdomen. Remove the salivary tracheal sacs in the abdomen.

Avoid the large sac in the abdomen which is not silvery in appearance, this is the *crop*. After you have picked off lateral thoracic sclerites, remove the dorso ventral muscles and the deltoid depressor of the trochanter. As you remove the dorsal longitudinal muscles of the mesothorax be particularly careful not to injure the gut which lies directly beneath these muscles. The living gut is translucent and is easily destroyed by careless use of forceps. Remove the tracheal twigs which holds the coils of the gut together. Uncoil them slowly. Trace the crop forward to the oesophagus. Hold the uncoiled gut in place with minuten pins and pour off the saline. Fix the gut with 70% alcohol or fixing fluid (Fig. 8.25).

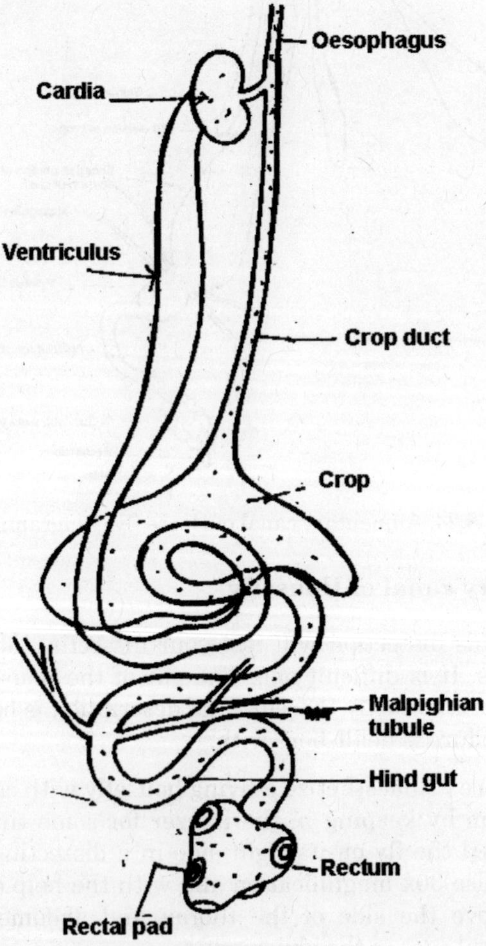

Fig. 8.25. Alimentary canal of Hosue fly

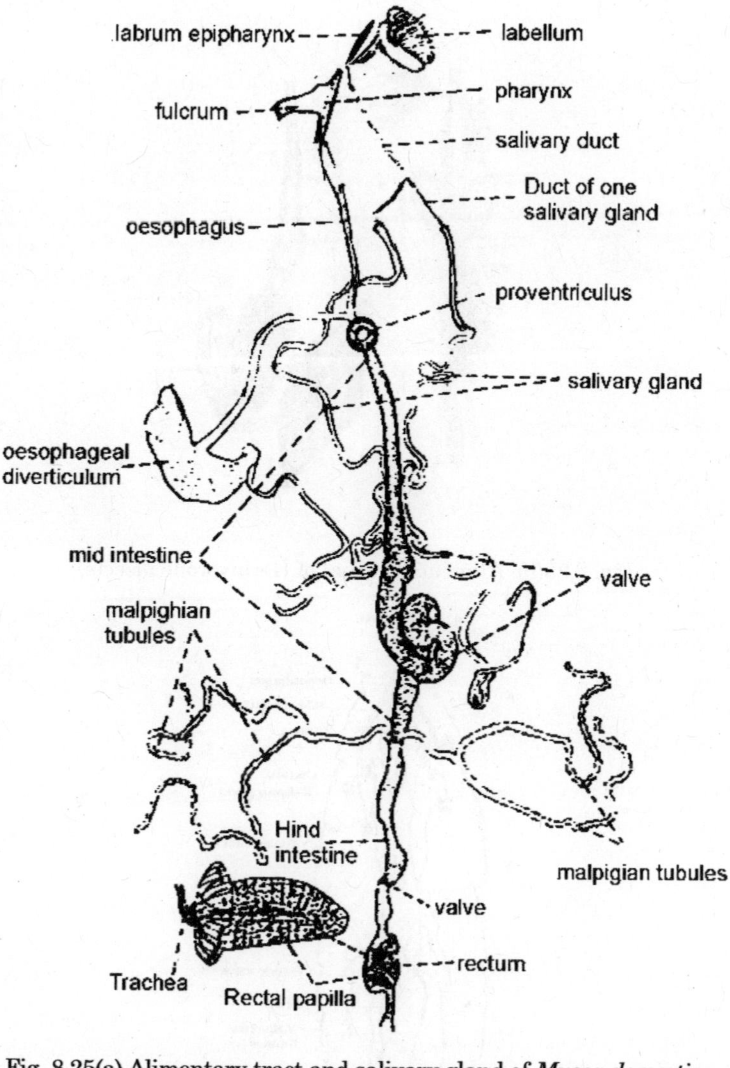

Fig. 8.25(a) Alimentary tract and salivary gland of *Musca domestica.*

(vi) Alimentary canal of grasshopper

Obtain alcohol preserved grasshoppers and prepare its dorsal dissection. Cut the wings and remove the tergal plates of the abdomen. It is similar to that of cockroach, (see the dissection of cockroach). Note the Pharynx, Oesophagus, Crop, Gizzard (Proventriculus), Mid gut and Hind gut (Fig. 8.26).

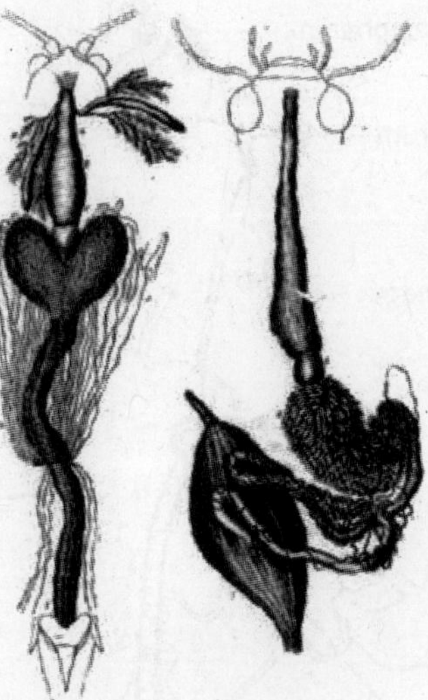

Fig. 8.25(b). Alimentary canal of Herbivorous insects.

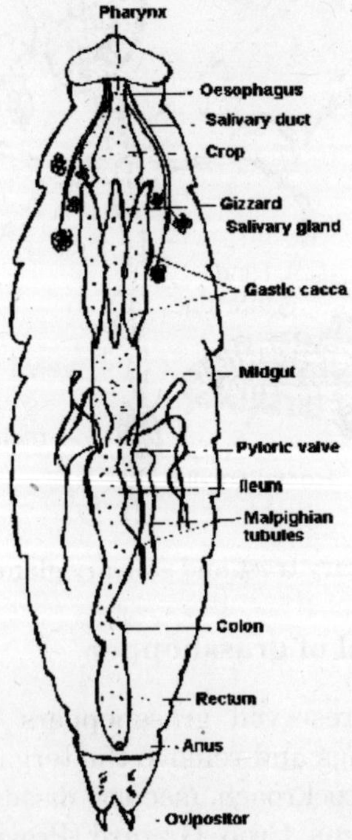

Fig. 8.26. Alimentary canal of Grasshopper.

(vii) Alimentary canal of cricket

Obtain the preserved insects or fresh ones. Prepare it for dorsal dissection. Remove the tergum from the abdomen and thorax and expose the different parts of the alimentary canal (see cockroach and grasshopper) (Fig. 8.27).

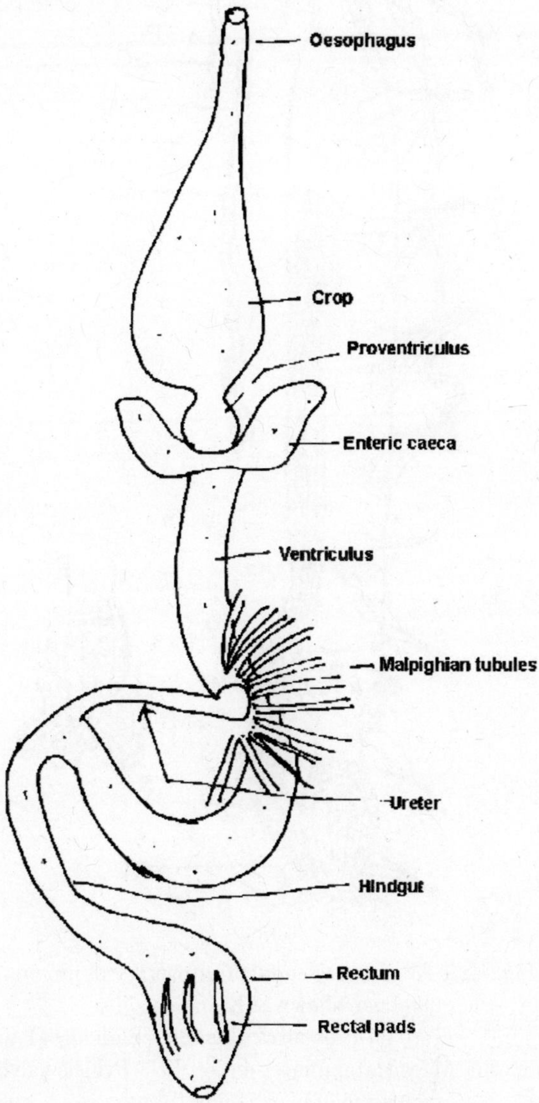

Fig. 8.27. Alimentary canal of cricket.

Since the basic plan of the dissections of all insects is similar therefore the description of following insects is not given here, only figures are given (Figs. 8.28, 8.29a & b, 8.30a & b, 8.31, 8.32, 8.33).

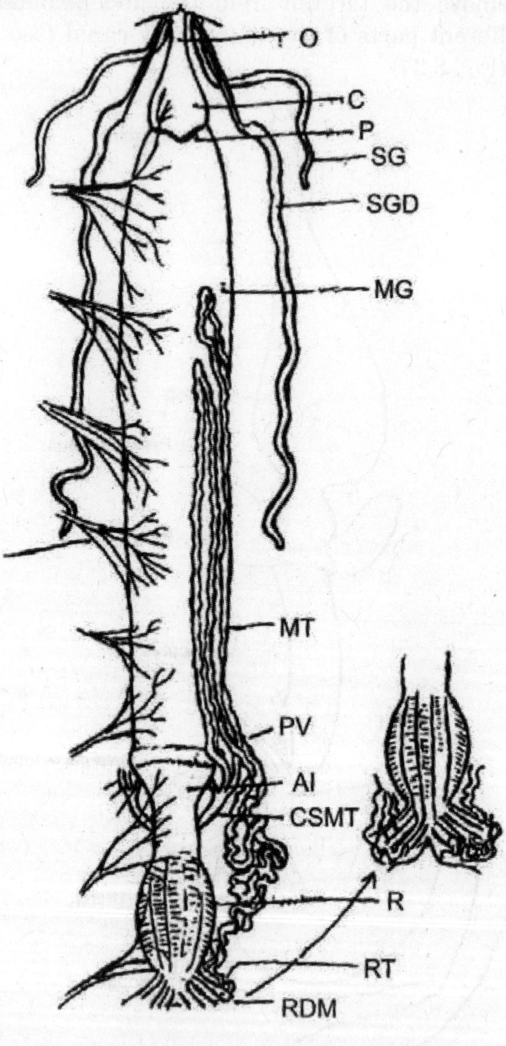

Fig. 8.28. Alimentary canal of caterpillar of junonia. Trcheae shown only on one side. (O - oesophagus, C - crop, P - Proventriculus, SG - Salivary gland, SGD - Silk gland, MG - Mid gut, MT - Malpighian tubules, PV - Pyloric valve, Al - Anterior intestine, CSMT - Common stem of malpighian tubules, R - Rectum, RT - Rectal tube, RDM - Rectal dilator muscles, TB - Tracheal bundle)

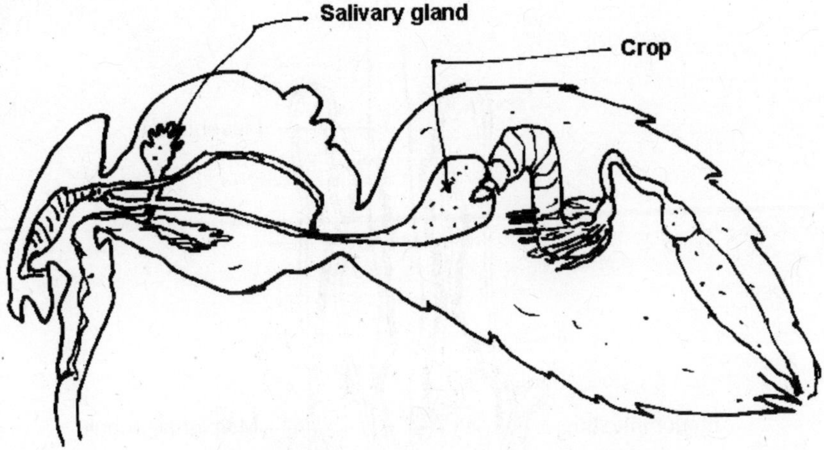

Fig. 8.29a. Alimentary canal of Honey bee (in situ)

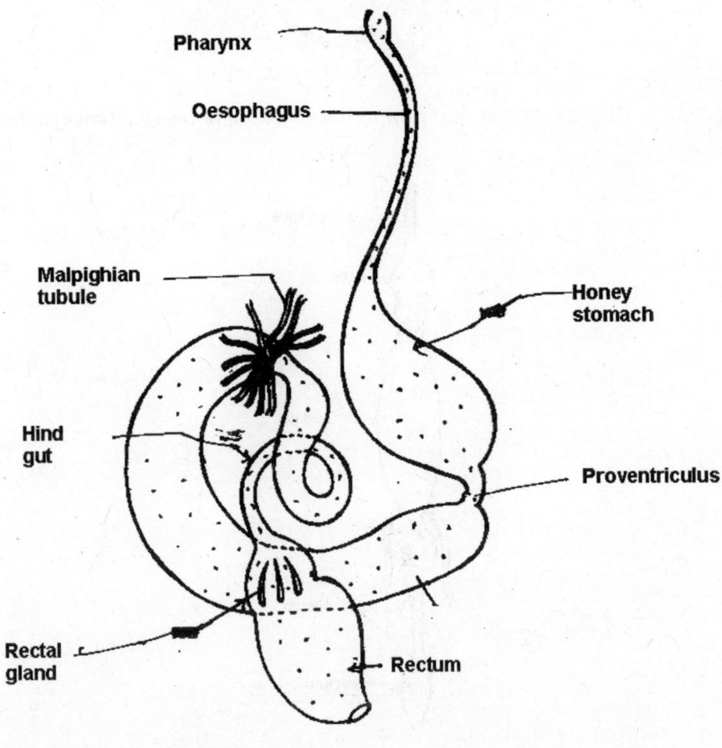

Fig. 8.29b. Alimentary canal of Honey bee (Diagrammatic)

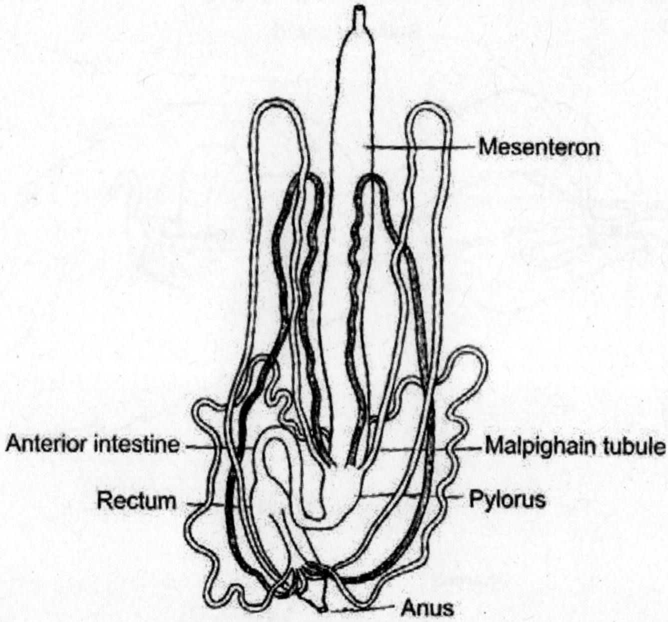

Fig. 8.30a. Digestive tract and malpighian system of beetle, *Tenebrio molitor* Larva

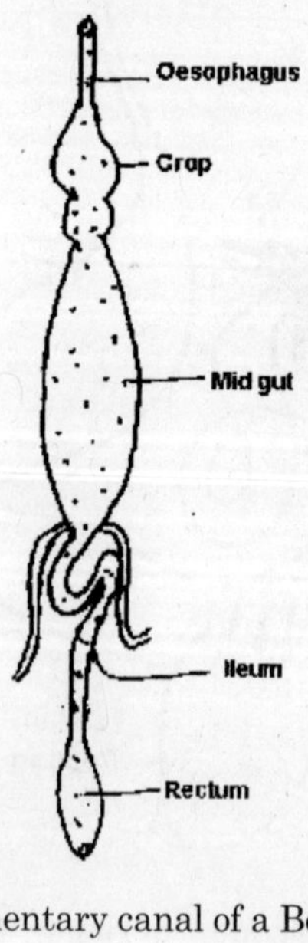

Fig. 8.30b. Alimentary canal of a Beetle (Generalised)

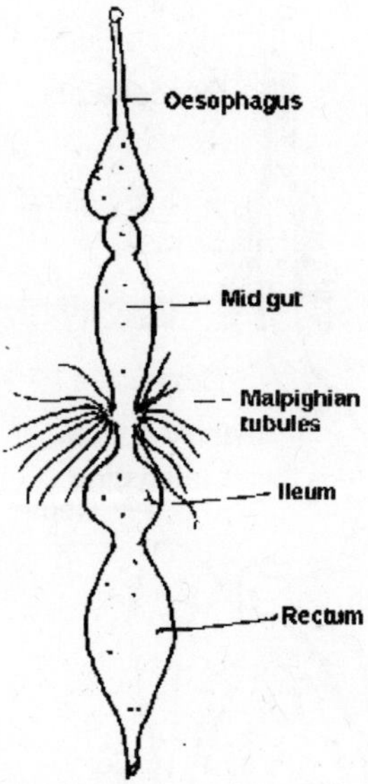

Fig. 8.31. Alimentary canal of dragon fly nymph.

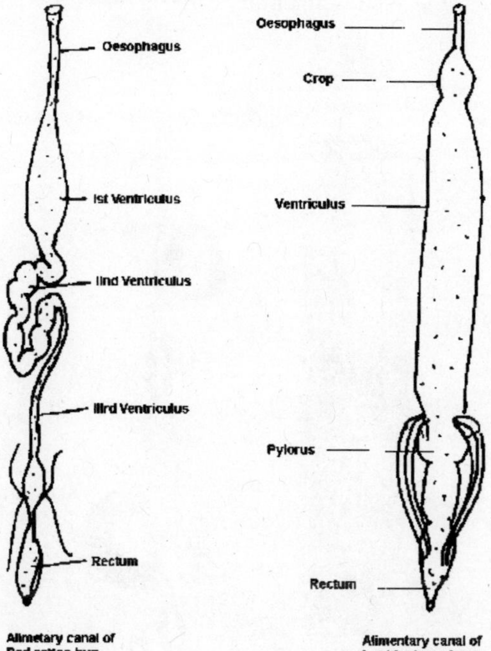

Fig. 8.32. Alimentary canal of Red cotton bug and Lepidopterous larva

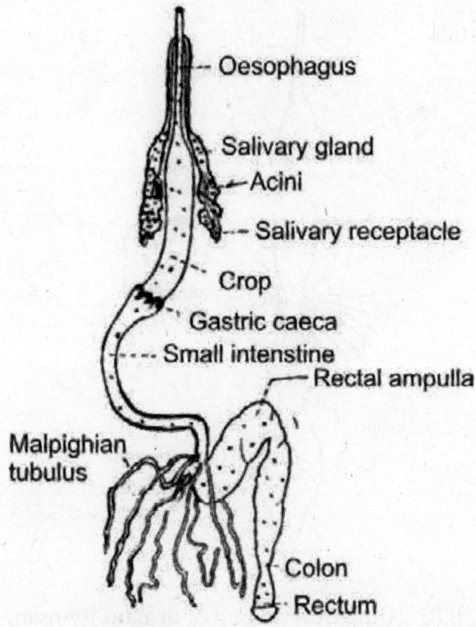

Fig. 8.33. Alimentary canal of termite.

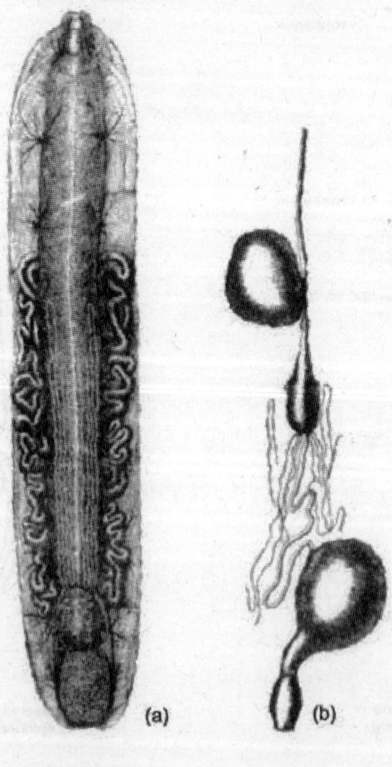

Fig. 8.33a. Alimentary canal of silkworm (a) caterpillar, (b) moth.

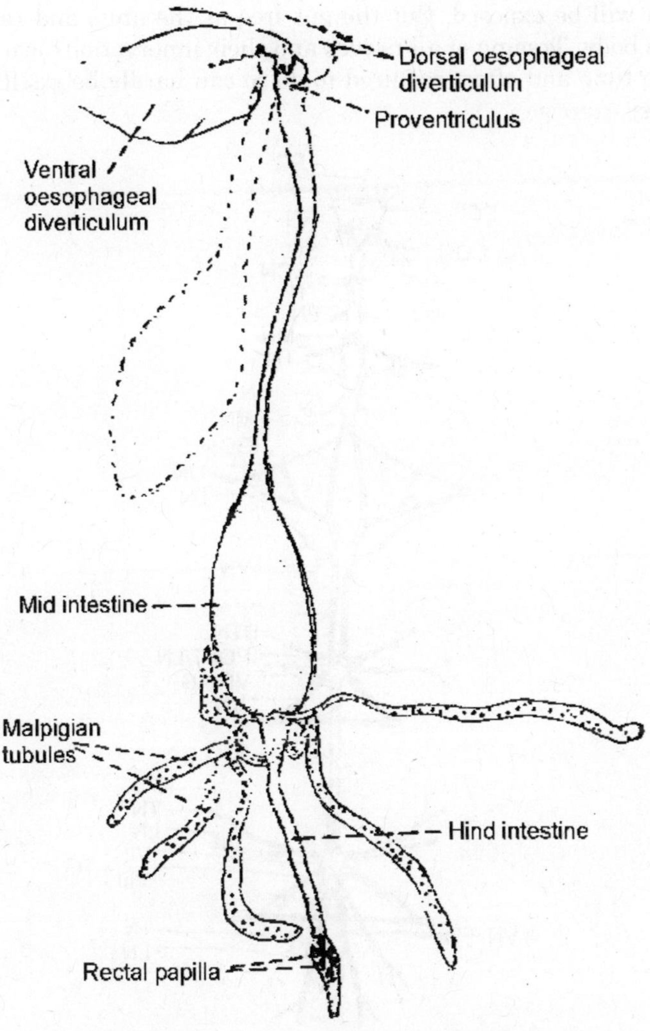

Fig. 8.33b. Mid and Hind intestine of Culex mosquito.

3. NERVOUS SYSTEM AND ENDOCRINE COMPLEX

(1) Nervous system of caterpillar (Junonia) (Fig. 8.34)

Prepare the caterpillar for dorsal dissection in a paraffin dish. Use stereoscopic dissecting binoculars. Cut the gut at the posterior end of the crop and slowly pull the gut. As you do this the ventral chain of ganglion will be exposed. Cut the gut free at the anus and remove it from the body. Tracing of all nerves and their innervations is a tedious task, the time and effort required to do so can hardly be justified in a laboratory exercise.

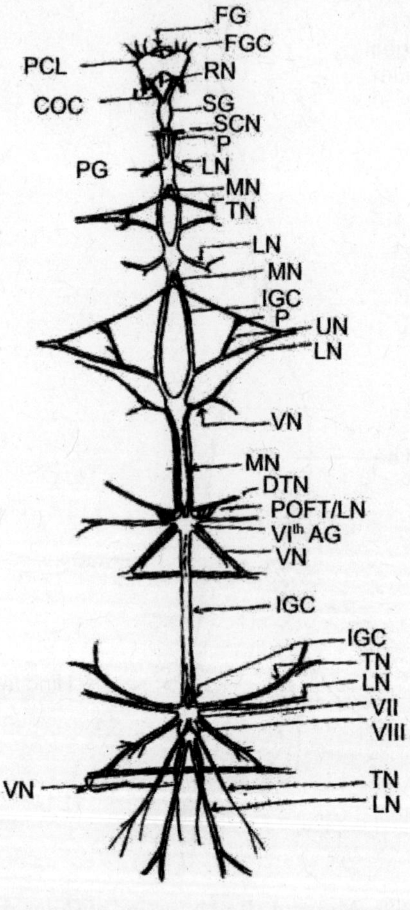

Fig. 8.34. Nervous system of Junonia (Butterfly)
(FG - Frontal ganglia, FGC - Frontal ganglionic connective, RN - Recurrent nerves, SG - Suboesophageal ganglion, SCN - Super connective nerve, P - Plexus, LN - Lateral nerve, MN - Median nerve, TN - Transverse nerve, IGC - Interganglionic connective, UN - Unitive nerve, VN - Ventral nerve, DTN-Dividing transverse nerve, POFT/LN - Plexus of transverse, lateral nerve, AG - Abdominal ganglion, TG - Terminal ganglia, PG - Prothoracic ganglia, COC - Circumoesophageal connective, PCL - Protocerebral lobe)

The central nervous system of the caterpillar consists of the brain, the suboesophageal ganglion, three thoracic ganglia and 8 abdominal ganglia. The peripheral nervous system consists of the cranial nerves and the nerves associated with the ventral chain of ganglia. The ganglia are joined to each other by longitudinal bundles of axons, called *connectives*. In some places these connectives are fused. Actually each apparent ganglia is a composite structure, composed of a pair of laterally fused ganglia. Internally, the fused ganglia are joined by bundles of short transverse fibres called *commissures*. Associated with the foregut is a series of nerves and ganglia, these arise from the brain and constitute the so called *stomodeal nervous system*. This division of the nervous system has also been called the *sympathetic nervous system* or the *Stomatogastric system*.

(a) Abdominal ganglia

Because the nervous system is easier to follow in the abdomen, we will start at the back of the caterpillar and work towards cephalic end.

(i) First six abdominal ganglia

The first 6 abdominal ganglia are located near the middle of their respective segments. The ganglia and their nerves are the same in each of these segments so that a description of one will serve for all. Each ganglion is connected to the succeeding ganglion by a pair of *interganglionic connectives*. A pair of *lateral nerves* project, one on each side, from the latero anterior margins of the ganglion. *Ventral nerves* also arise to supply the ventral area of the segment. They run parallel to a pair of trachea. Short *median nerves* arise from interganglionic connectives. Median nerves produces two diverging *transverse nerves*. Just lateral to the ganglion, the transverse nerves and the lateral nerves are joined by a web of *plexus* of nerves fibres. Beyond the plexus, the transverse nerves diverge from the lateral nerves in a cephalo lateral direction.

(ii) Seventh and eighth abdominal ganglia

7th and 8th abdominal ganglia are the terminal ganglia of the abdomen and are situated in the 7th and 8th segments. Interganglionic connectives between them is very short and they seldom can be distinguished externally. It bears a pair of lateral nerves and a pair of ventral nerves. The interganglionic connective leading to 7th ganglion produces a median nerve which in turn divides into a pair of transverse nerves. A *plexus* is also present which joins the transverse and lateral nerves on each side. The eighth ganglion is composite one. Find the short median nerve which arises dorsally at the junction of the seventh and eighth ganglia. The relatively large *lateral nerves* arise from the posterior end of the eighth abdominal ganglia; they pass well into the eighth segment before branching. A pair of *ventral nerves* leaves the posterior end of the eighth ganglia.

(b) Thoracic and cephalic ganglia

There are three thoracic ganglia which corresponds to the three thoracic segments. It will be necessary at this time to crack the head open along the ecdysial line and to trim the foregut back to the brain. Before removing the crop however, note the delicate nerve running along the dorsal median line of the foregut. This is the *recurrent nerve* which belongs to the *stomodeal nervous system*.

Mesothoracic and metathoracic ganglia and nerves are similar therefore description of metathorax will serve for both. It is required to tease away some of the fat and remove some of the ventral muscles to follow the nerves. It requires patience as it is difficult one. Find the first abdominal ganglion. It is close to metathoracic gang-lion because the connectives are short. A pair of long interganglionic connectives join the meta and mesothoracic ganglia; these connectives are widely separated for most of their length but converge as they approach the ganglia. The lateral nerves do not leave the metathoracic ganglion directly but, instead, follow the interganglionic connectives for a short distance before they diverge and independently pass off in a latero-cephalic direction. A pair of ventral nerves leave the lateral surface of the ganglion. The short median nerve leaves the posterior end of *mesothoracic ganglion* instead of being attached to the connectives. It produces a pair of *transverse nerves* which join the lateral nerves at the sides of the metathoracic segment. Note that the lateral nerves from the metathoracic ganglion are also joined to the transverse nerve coming from the metathoracic ganglion by a pair of *unitive nerves*. Associated with the unitive nerves is a *plexus*. A similar relationship exists between the lateral nerves of the mesothoracic segment and the transverse nerves issuing from the prothoracic ganglion.

The *prothoracic ganglion* is joined to the suboesophageal ganglion by a pair of relatively short interganglionic connectives. The median nerves leaves the posterior end of the prothoracic ganglion. The most conspicuous pair of nerves associated with the prothoracic ganglion are the lateral nerves. The prothoracic ganglion is not preceded by the usual median and transverse nerves instead the interganglionic connectives joining the suboesophageal and prothoracic ganglia are joined to a transverse *supra connective nerve* by a plexus of small fibres.

Suboesophageal ganglion and brain are situated in the head.

Gently lift the remains of the foregut without damaging the brain. Find the *suboesophageal ganglion.* This give rise to *mandibular, maxillary* and *labial nerves.* Two large connectives extend from the anterior end of the suboesophageal ganglion, these are the *circumoesophageal connectives* which pass dorsad, one on each side of the gut, to the brain.

The brain lies above the gut, the two large tubes seen in the dorsal dissection are the *protocerebral* lobes. Several nerves leaves the anterior, ventral surface of the brain, these supply the optic, antennal and clypeolabral organs. Immediately anterior to the median part of the brain are a number of muscle fibres which serve the dorsal dilators of the pharynx. Push these back and locate the short *frontal ganglion connectives* which join the brain to a small median ganglion. This lies on the dorsal surface of the gut and is called *frontal ganglion.* The delicate recurrent nerve which follows the median line of the crop, and which was noted earlier in this exercise, proceeds posteriorly from the frontal ganglion. The frontal ganglion, the frontal ganglion connective and the recurrent nerve are the part of the *stomodeal nervous system*.

(2) Nervous system of adult Horse Fly (*Tabanus*)

Obtain a preserved female horsefly (*Tabanus*). Remove the gut except for that portion of the fore gut that passess beneath the brain. Find the large *thoracic ganglionic mass* in the anterior third of the thorax; this composite ganglion represents the fused ganglia of the pro, meso and metathoracic ganglia and the first abdominal segment. A stout interganglionic connective extends cephalad from the anterior end of the thoracic ganglionic mass; it passess through the cervix and attaches to the suboesophageal ganglion. In the cervix, the connective produces two pairs of *cervical nerves*.

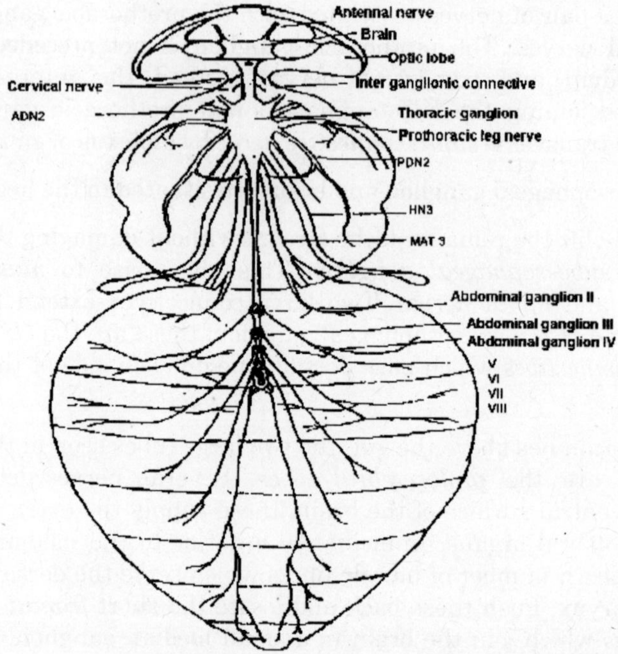

Fig. 8.35. Nervous system of Horse fly (PDN2 - Posterior dorsal meso thoracic nerve, HN3 - Hatere nerve, MAT3 - Metathoracic and abdominal nerve, ADN2 - Anterior dorsal mesothoracic nerve)

A nerve of large nerve trunks leave the thoracic ganglionic mass. The following can be traced and uncovered easily. Three thoracic ganglia and one abdominal ganglia are involved in the thoracic mass. Although the nerve trunks are paired, only one of each is being described here.

The anterior dorsal nerve (ADN_1)

It arises from the anterior, dorsal margins of the thoracic mass and extends anterolateral. It branches, a short distance from the ganglion, to form nerves that supply certain cervical and prothoracic muscles.

The prothoracic leg nerve (LN_1)

It arises from the ganglion directly beneath the proceeding nerve. It is stout and supplies the prothoracic legs.

The anterior dorsal mesothoracic nerve (ADN_2)

It is stout and arises posterior to (ADN_1), that produces numerous branches innervating flight muscles of the mesothorax.

The posterior dorsal mesothoracic nerve (PDN_2)

It arises from thoracic ganglionic mass just dorsal to ADN_2, then passes laterocaudad. A short distance beyond the ganglionic mass, it sends of a branch that curves back to the mid line of the thorax; this branch innervates the great, dorsal longitudinal muscle of the mesothorax. The rest of the nerve passes into the posterior part of the mesothorax.

The haltere nerve (HN_3)

It is a long, ribbon like that leaves the posterolateral edge of the ganglionic mass and extends to the halteres.

The metathoracic and abdominal nerve

Metathoracic and abdominal nerves are arised from the posterior end of the ganglionic mass, just lateral to the interganglionic connective. This large nerve trunk produce branches that supply haltere muscle and muscles of hind legs. Notice also that one of its branch passes into the abdomen and innervates the first abdominal segment. The *interganglionic connective* that leaves the posterior end of the thoracic ganglionic mass, follows the mid ventral line of the body into the abdomen; it joins the thoracic ganglionic mass to the second abdominal ganglion.

Now return to head and expose the brain. The optic lobes join the protocerebrum of the brain to the compound eyes. The *antennal nerve* project anteriorly from the brain. The *suboesophageal ganglion* lies directly beneath the brain. The *circumoesophageal connectives* are very short and stocky, leaving only narrow passageway between the brain and suboesophageal ganglion. The supra and suboesophageal ganglia appear to have fused into a compact mass.

The ganglia that lie in the abdomen are 2nd to 8th. The ganglionic connectives are very short and thus the central nervous system of abdomen is restricted to the first two segments. The last three ganglia are fused due to the shortening of the connectives. The ganglia can be easily identified by following their nerves to the respective abdominal segments. Each ganglion supplies one pair of nerve trunks which ultimately branch to form a number of nerves.

The second abdominal ganglion (II) lies at the anterior end of the abdomen, it produces a pair of nerve trunks which innervates the second abdominal segment. The inter ganglionic connective between ganglion II and III is relatively short, but is still the longest connective in the abdomen. Trace the pair of nerves from the third abdominal ganglion (III) to the third abdominal segment. The 4th abdominal ganglion (IV) is discrete although the connectives that join it to the third and fifth ganglion are extremely short. The 5th, 6th, 7th and 8th abdominal ganglia are so closely associated that they form a large oval mass. The nerves leaving this oval mass define the constituent ganglia. The nerves leaving 5th through 8th abdmoinal ganglia usually fuse for a part of their lengths. For this reason two exceedingly robust nerve trunks leave the posterior ganglionic mass of the abdomen. Ultimately each segmental nerve leaves the composite nerve trunk to pass to it respective segment.

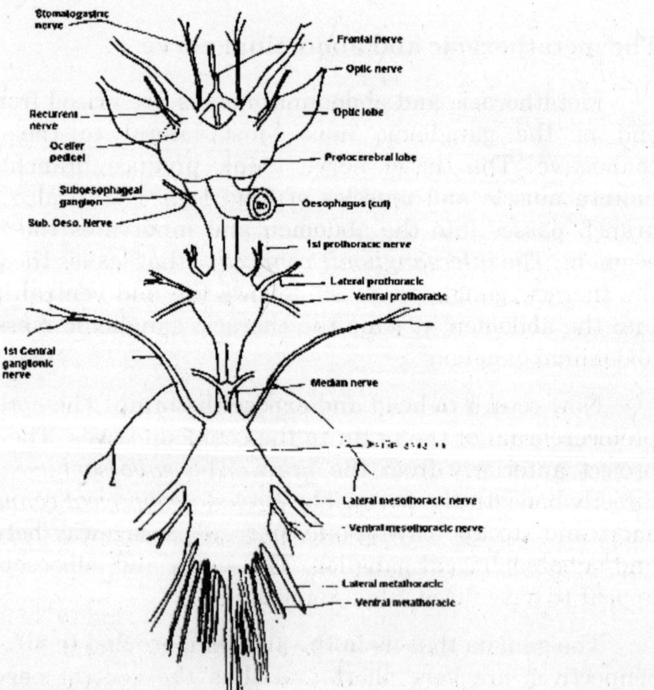

Fig. 8.36. Nervous system of milkweed bug (Dorsal view)

3. Nervous system of a Bug *(Oncopeltus)*

Prepare the insect for dorsal dissection in paraffin dish. Use dissecting binoculars. Look at the figure and trace the nervous system accordingly (Fig. 8.36).

4. Nervous system of Belostoma

Prepare the insect for dorsal dissection in a paraffin dish. Use dissecting binoculars. Look at the figure and trace the nervous system accordingly (Fig. 8.37).

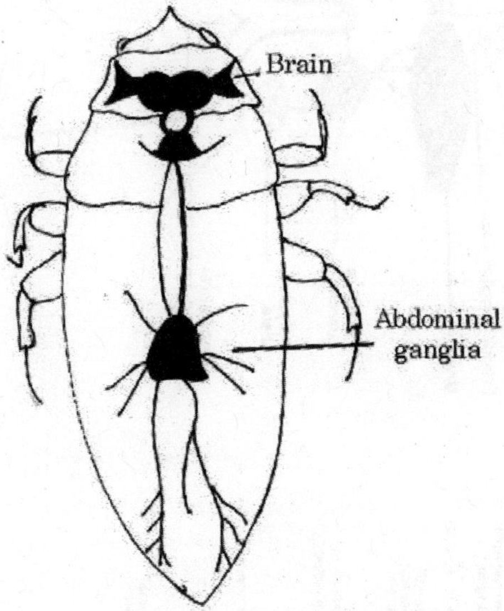

Fig. 8.37. Nervous system of Belostoma.

5. Nervous system and Endocrine complex of Cockroach and Grass hopper

Prepare the insects for dorsal dissection in a paraffin dish. Use dissecting binoculars. See the figure and trace the nervous system (Fig. 8.38 and 8.38a).

6. Nervous system of Honeybee

Imbed the bee in paraffin dish. Use dissecting binoculars. See the figure and trace the nervous system (Fig. 8.39 and 8.40a).

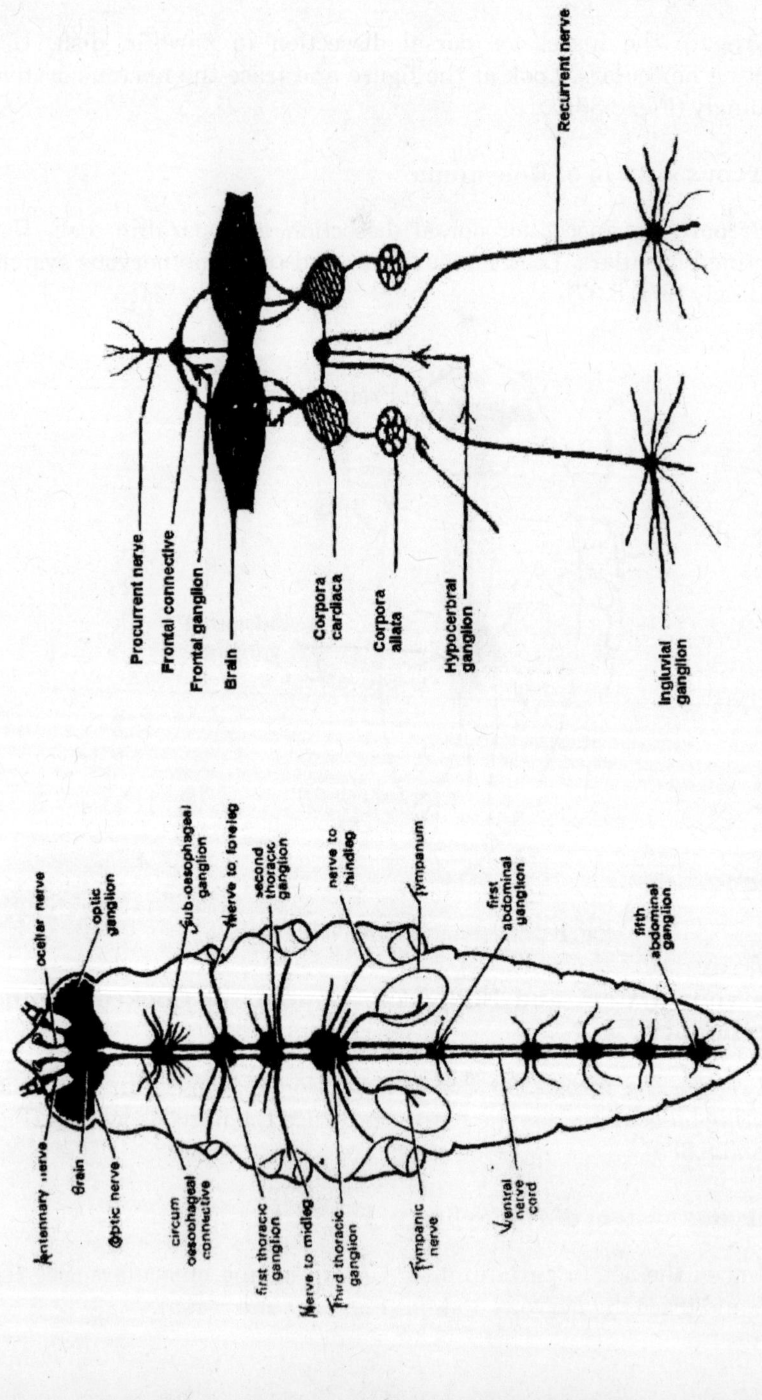

Fig. 8.38. Nervous system of Grasshopper (in situ) and stomodeal sympathetic nervous system (cockroach)

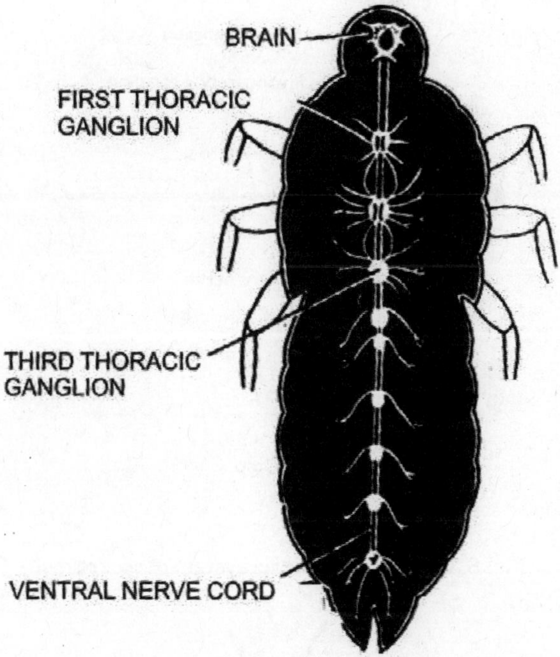

Fig. 8.38b. Nervous symptom of Cockroach.

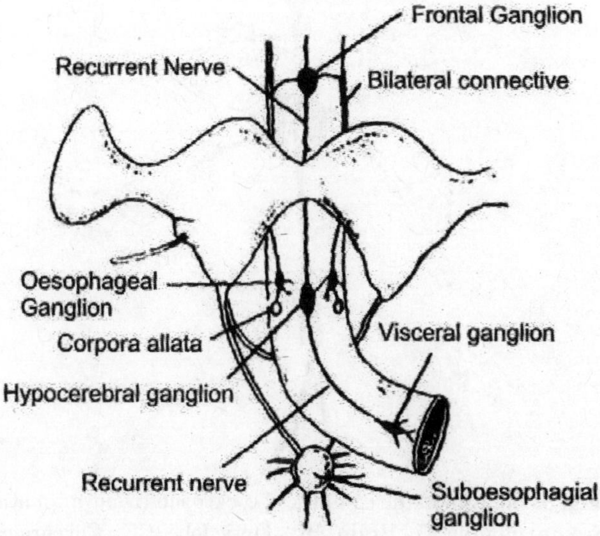

Fig. 8.38c. Brain of Cockroach showing visceral nervous syst.

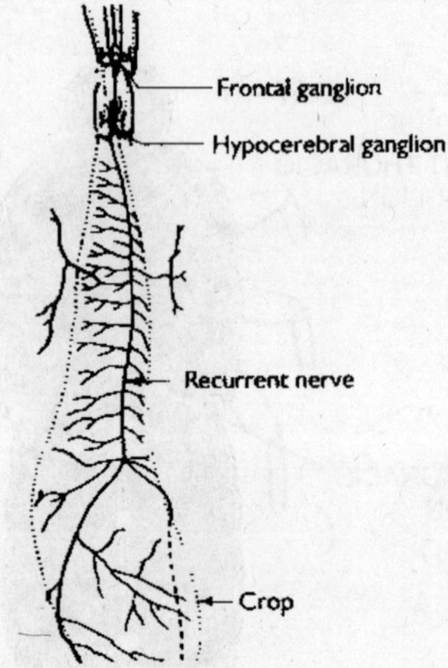

Fig. 8.38d. Stomatogastric system of cockroach.

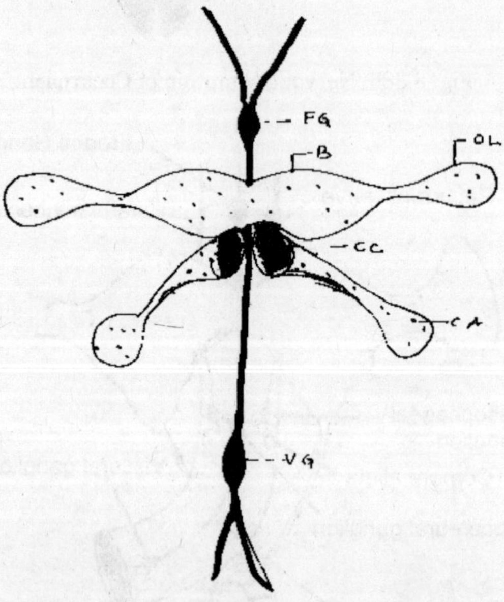

Fig. 8.38e. Endocrine complex of cockroach (Diagrammatic)
(FG - Frontal ganglia, B - Brain, OL - Optic lobe, CC - Corpora cardiaca, CA - Corpora allata, VG - Visceral ganglia)

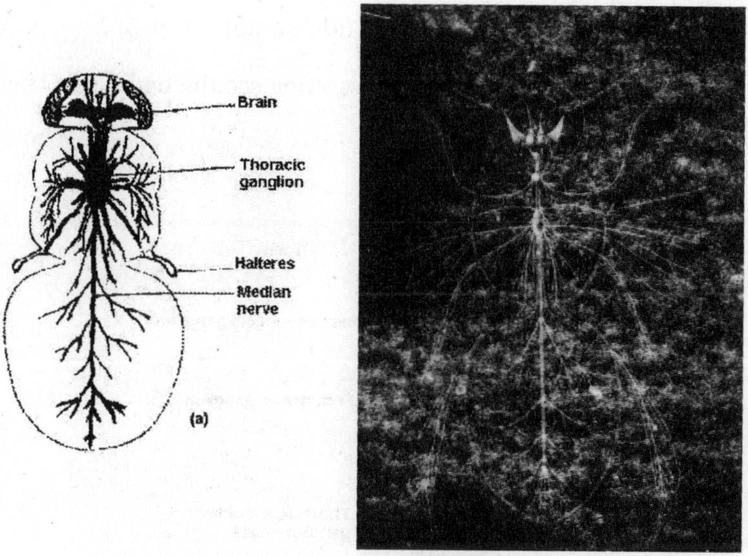

Fig. 8.39. Nervous system of Honey bee (dissected).
a. Nervous system of Honey bee *(in situ)*

7. Nervous system of Housefly (*Musca*)

Imbed the fly is paraffin dish. Use dissecting binoculars. See the figure and trace the nervous system (Fig. 8.40).

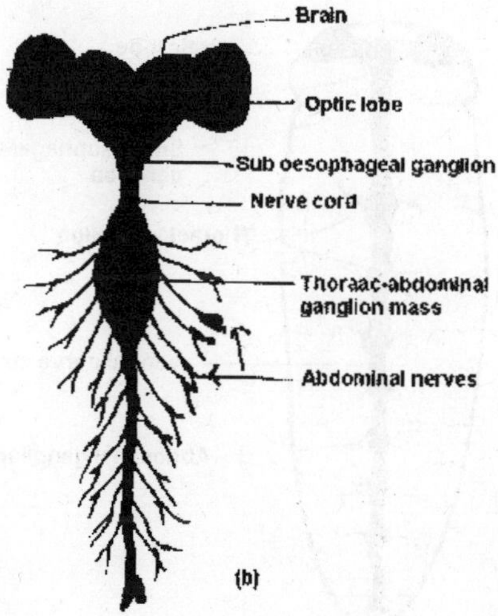

Fig. 8.40. Nervous system of House fly (Diagrammatic)

8. Nervous system of plant bug and cricket

Prepare the insect for dorsal dissection see the figure and trace the nervous system (Fig. 8.41).

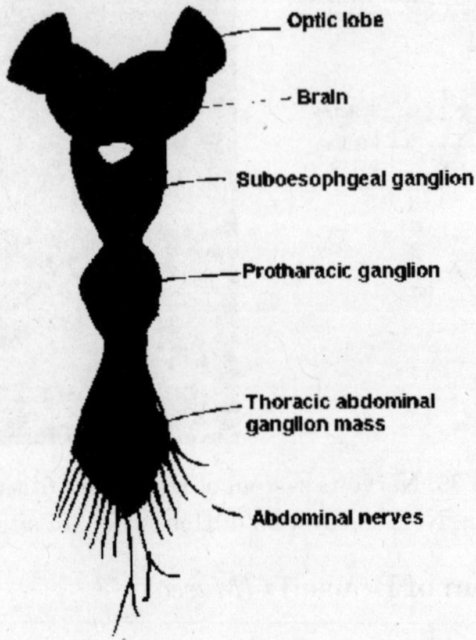

Fig. 8.41. Nervous system of plant bug

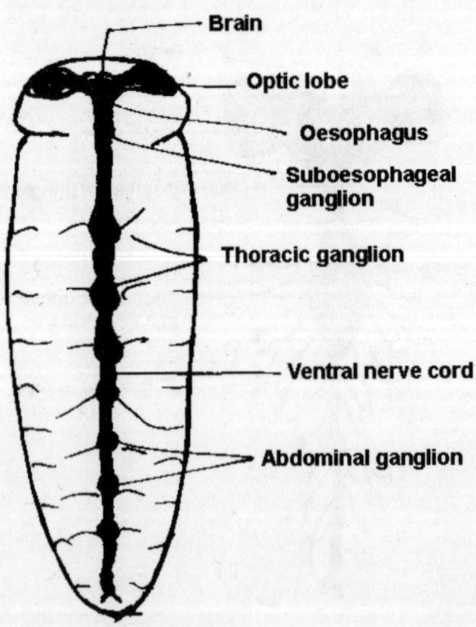

Fig. 8.42. Cricket central nervous system (dissected)

9. Nervos system of Grass hopper, Cockroach, Cricket, Water beetle and silkworm

Prepare the insect for dorsal dissection, see the figure and trace the nervous system.

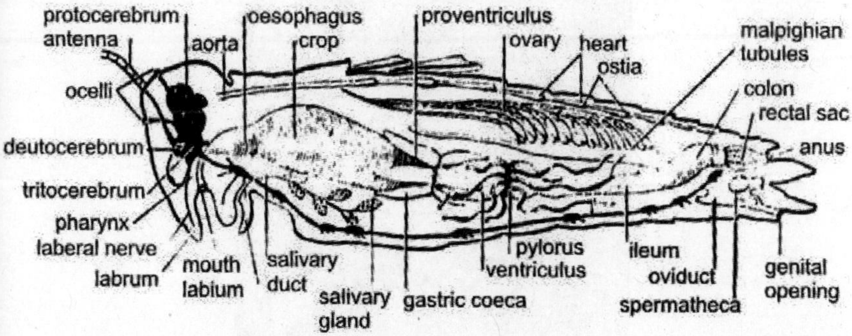

Fig. 8.42(a). Internal anatomy of Grasshopper showing nervous system, alimentary canal and reproductive system.

10. Dissection of Grasshopper

Showing internal anatomy.

11. Tracheal system of Grasshopper

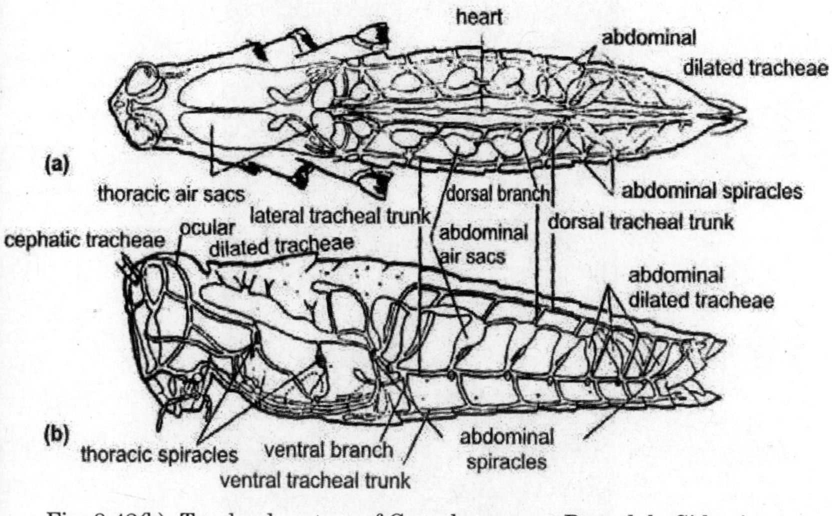

Fig. 8.42(b). Tracheal system of Grasshopper, a. Dorsal, b. Side view.

12. Nervous system of Water Beetle

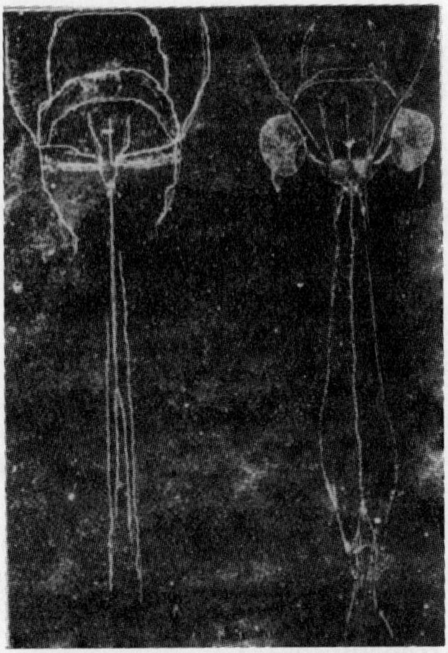

Fig. 8.42(c). (a) larva (b) adult

13. Nervous system of Silk Worm (Adult and Larva)

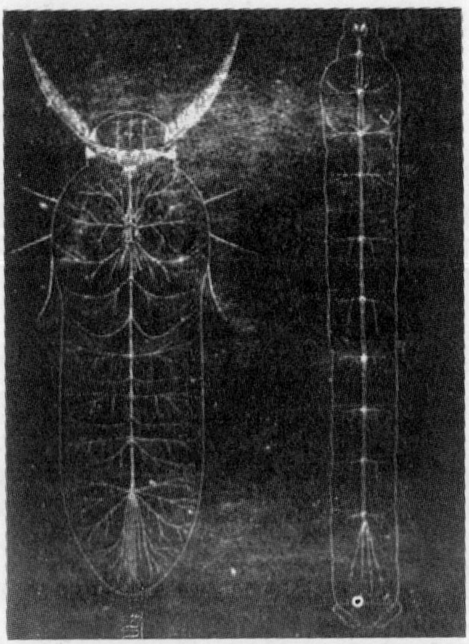

Fig. 8.42(d). Nervous system of Silk worm.

4. REPRODUCTIVE SYSTEM

Since the general plan of reproductive system is common to all insects therefore the individual description are not being given. See the figures and trace the reproductive systems of the given insects.

(i) Reproductive system of Grasshopper

Prepare the grasshopper for dorsal dissection in a paraffin dish. Cut the lateral membranes between terga and sterna and also remove the tergal plates of abdomen and thoracic region. Carefully examine and expose the male and female reproductive systems. Draw labeled diagram of your dissection in the practical record.

Note that the *Male reproductive system* consists **testes** (4, 5 & 6th abdominal segment) **vasadeferrntia, ejaculatory** duct, **aedegus** and **accessory glands**. The *female reproductive system* comprises of **ovaries**, **oviduct, vagina spermatheca** and **accessory glands.**

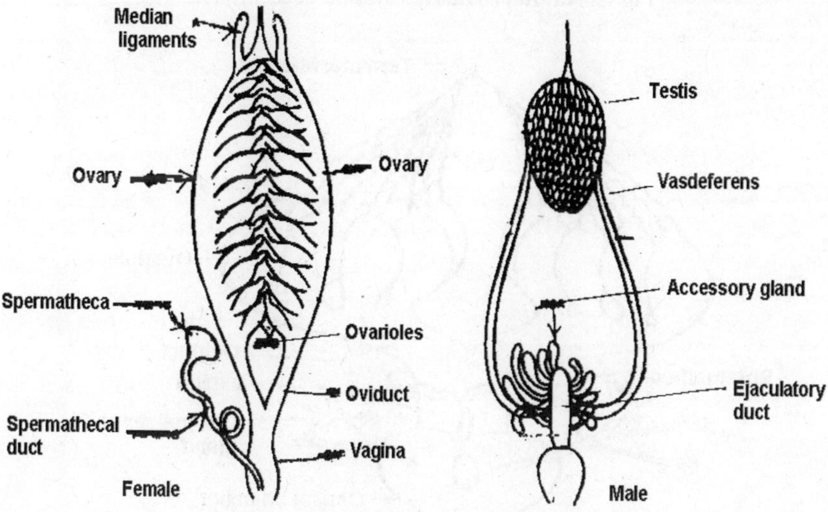

Fig. 8.43. Reproductive system of Grasshopper.

(ii) Reproductive system of Cockroach

Similar to grasshopper. See the figure and trace it accordingly (Fig. 8.44a & b).

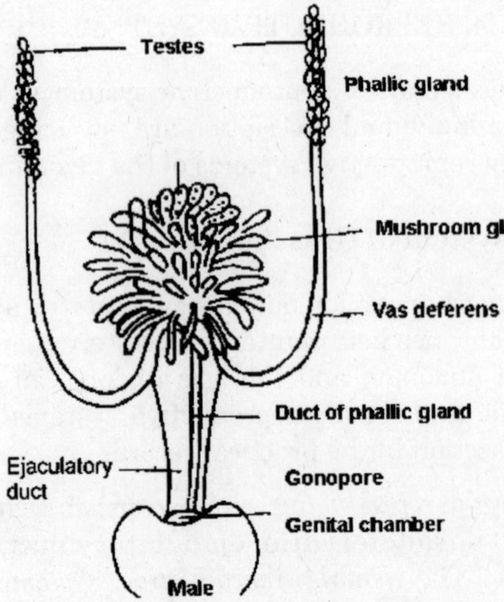

Fig. 8.44a. Reproductive system of cockroack.

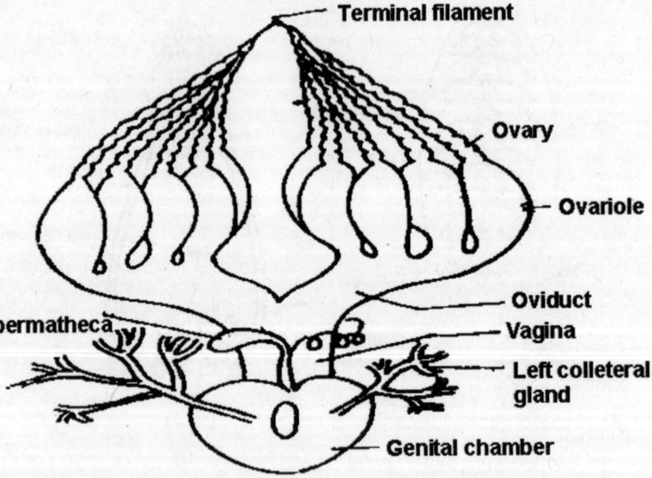

Fig.8.44b. Reproductive system of cockroach (Female)

(iii) Reproductive system of Dragon fly

Prepare the insect for dorsal dissection in a paraffin dish. Use dissecting binoculars and trace the reproductive systems (Fig. 8.45a & b).

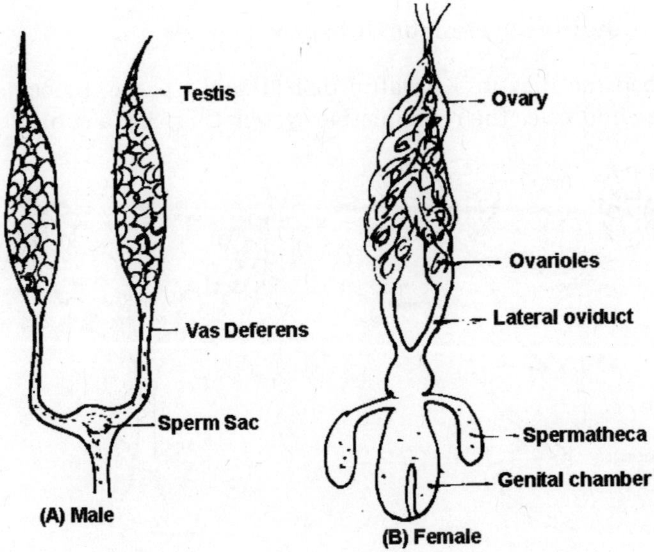

Fig. 8.45. Reproductive system of Dragon fly.

(iv) Reproductive system of Lepidoptera

Prepare the insect for dorsal dissection in a paraffin dish. Use dissecting binoculars. See the figure and trace the reproductive system (Fig. 8.46a & b).

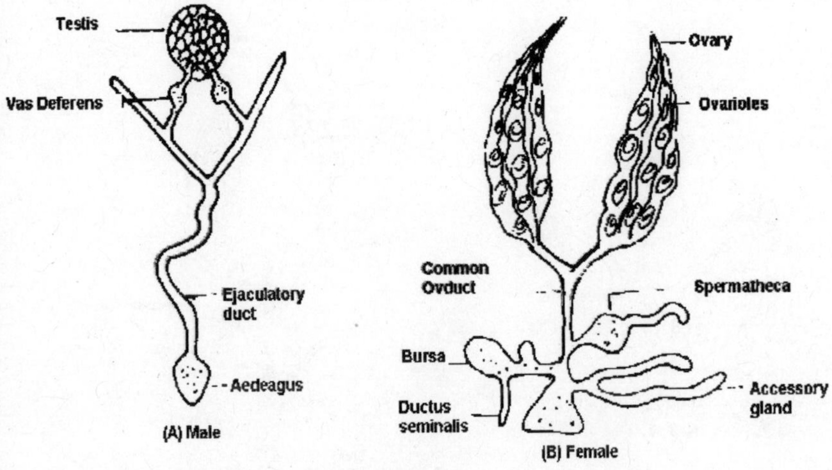

Fig. 8.46. Reproductive system of Lepidoptera

(v) Reproductive system of Housefly

Imbed the flies in a paraffin dish. Use dissecting binoculars. See the figure and trace the reproductive system (Fig. 8.47a & b).

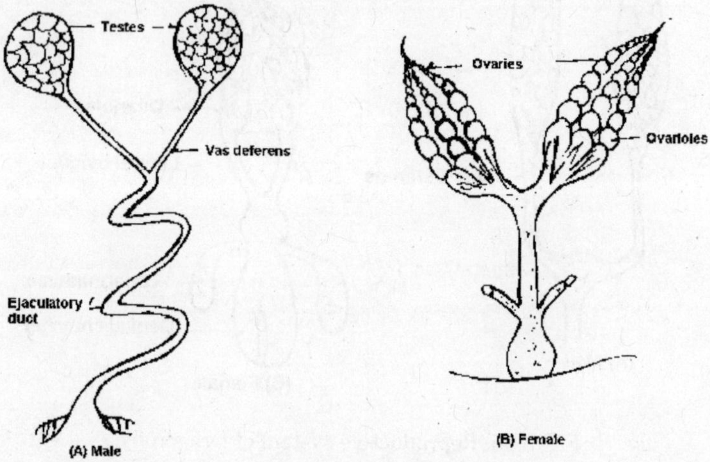

Fig. 8.47. Reproductive system of Musca.

(vi) Reproductive system of Honey bee

Imbed the bee in paraffin dish. Use dissecting binoculars. See the the figure and trace the reproductive system (Fig. 8.48a & b).

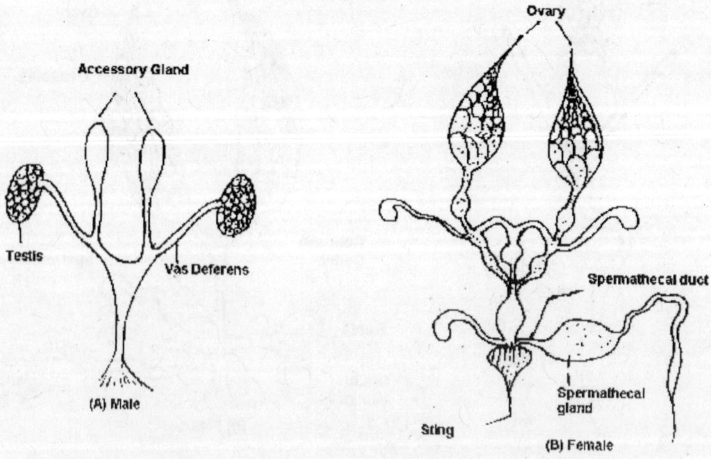

Fig. 8.48. Reproductive system of Honey Bee.

(vii) Reproductive system of Beetle

Prepare the insect for dorsal dissection is a paraffin dish. Remove the tergal plates and plurites. Keep aside the alimentry canal and trace the reproductive system (Fig. 8.49a & b).

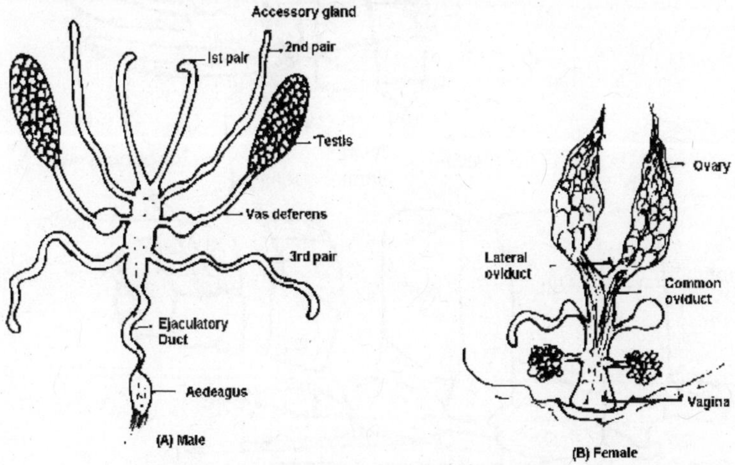

Fig. 8.49. Reproductive system of a Beetle.

(viii) Structure of male and female Genitelia (Generalized)

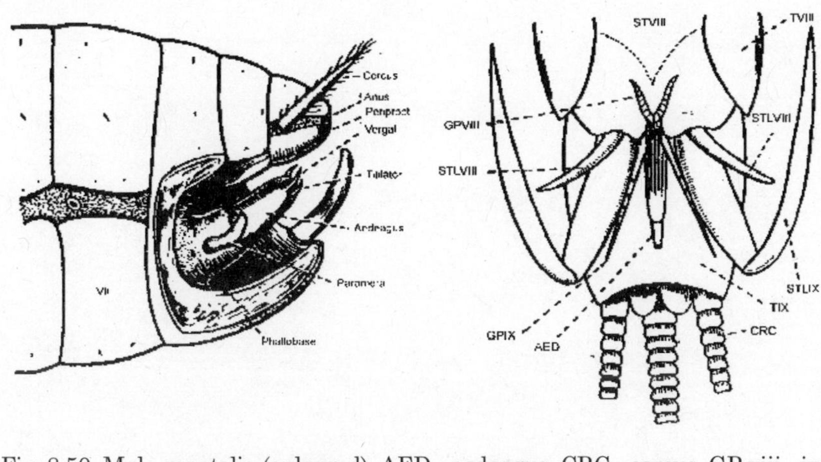

Fig. 8.50. Male genetelia (enlarged), AED - aedeagus, CRC - carcus, GP viii - ix - gonapophysis, Tix - tergum, STL - styles.

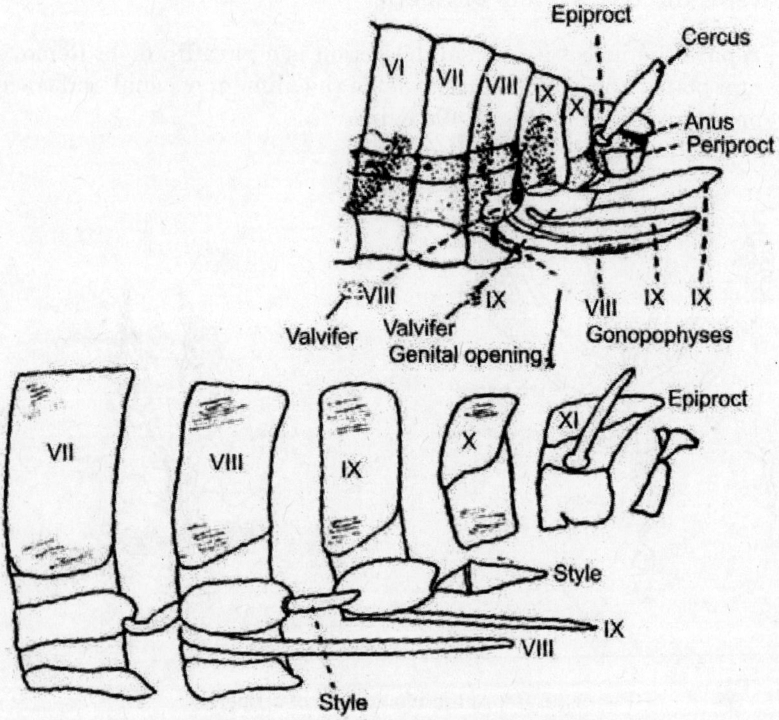

Fig. 8.51. Female genetelia (enlarged)

CHAPTER - 9

HISTOLOGICAL SLIDES

1. T.S. of midgut

1. The outer most layer is made of connective tissues below which lies the circular muscles.
2. Below circular muslce layer is a wavy epithelium lining which consist of two types of cells-
   (a) Tall columnar with microvilli forming a striated border binding the lumen. Function - Enzyme secretion and absorption.
   (b) Regeneration or nidicells situated at the base which replace torn out columnar epithelium cells.

(3) A delicate peritropic membrane is present in the central region which provides protection to the mid gut epithelium from hard particles of food.

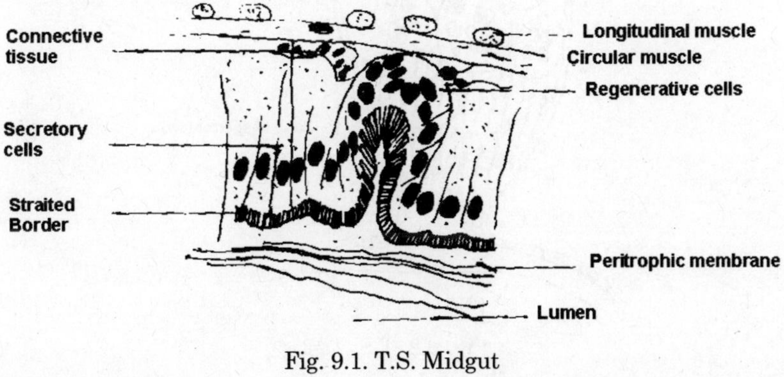

Fig. 9.1. T.S. Midgut

2. T.S. of testis

1. Testis is made up of series of tubular follicles.
2. Each follicle consist of a layer of epithelium resting on basement membrane.
3. Follicles are enclosed in a connective tissue sheath.
4. Each follicle contains following zones in succession -
   a. Zone of spermatocytes.
   b. Zone of maturation and reduction possess spermatids.
   c. Zone of transformation possess spermatozoa.

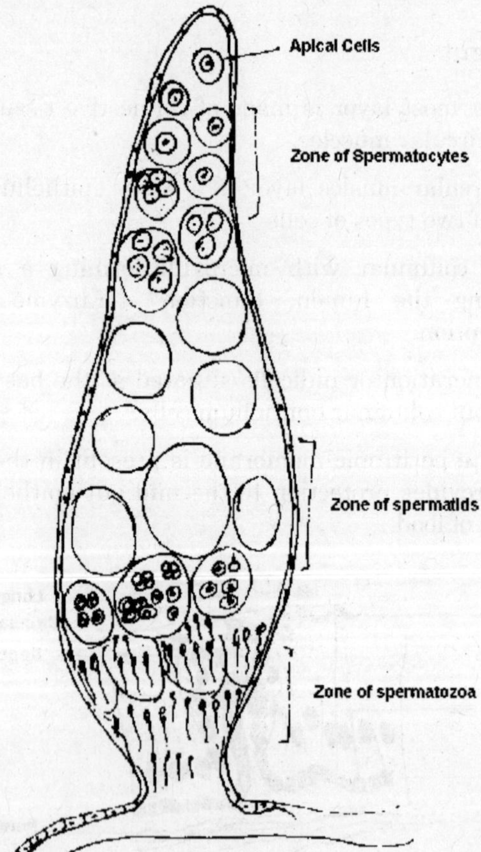

Fig. 9.2. T.S. Testis

3. L.S. of ovary

1. Each ovary is made up of series of egg tubules or ovarioles.
2. The ovariole consist of a chain of developing epithelium resting on a basement membrane.
3. It is divided into following zones —
   a. Terminal filament : It attaches all the ovarioles to the body wall.
   b. Germarium : Packed with primordial germ cells or oogonia. They become differentiate into oocytes and nutritive or nurse cells.
   c. Vitellarium : Consist of a series of oocytes, progressively larger towards the lower end.
   d. Ovariole stalk : They lead to oviduct.

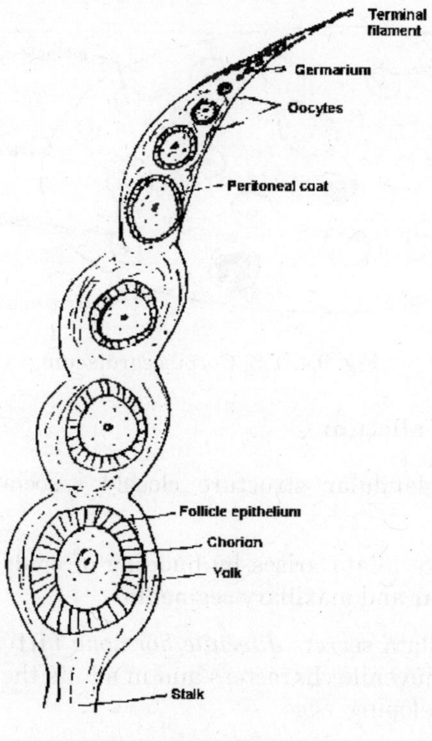

Fig. 9.3. L.S. of Ovary.

Panoistic ovary

Nutritive or nurse cells are absent. The yolk of egg is formed by the ephithelium of the egg follicles.

4. T.S. Corpus-cardiacum

Following cells are visible -

a. The distended bulbous ending of the axons from the neuro-secretory cells of the protocerebrum arising by way of two nerves. These endings are filled with neurosecretory granules.

b. Intrinsic cells of neurons with axon like process contains pale granules.

c. Interstitial cells covering nervous elements and with promi- nent nucleus. They liberate the neurosecretory product from the brain and also produce the neurosecretory product of their own.

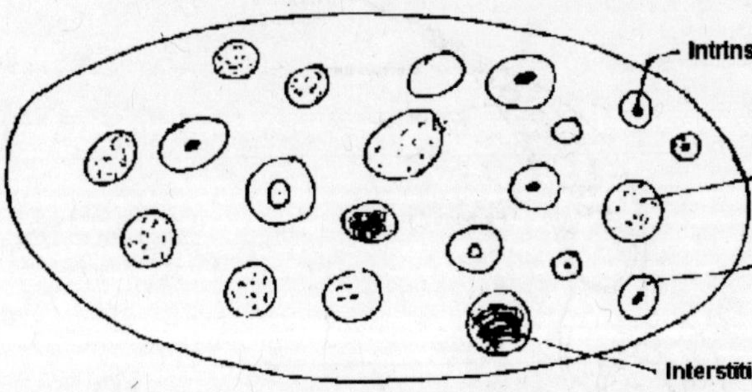

Fig. 9.4. T.S. Corpus cardiacum.

5. T.S. Corpum allatum

1. It is a glandular structure closely associated with corpora cardiaca.
2. The corpora allata arises by budding of epidermal cells betwen mandibular and maxillary segments.
3. Corpora allata secrets *Juvenile hormona* (JH) or *neotenin* which maintain juvenile characters and in adults the deposition of yolk in the developing eggs.

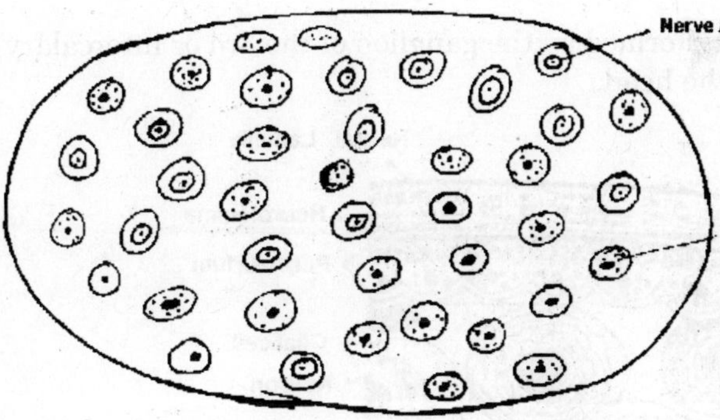

Fig. 9.5. T.S. *Corpus allatum*

6. T.S. thoracic ganglion

1. The perikarya of motor and internuncial neurons are aggregated to form ganglion within which they are grouped peripherally.
2. The centre of the ganglion is occupied by the neuropile, a complex of afferent internuncial and efferent fibres of their supporting glial elements.
3. Within the neuropile, groups of fibres oriented so as to form fibre tract.

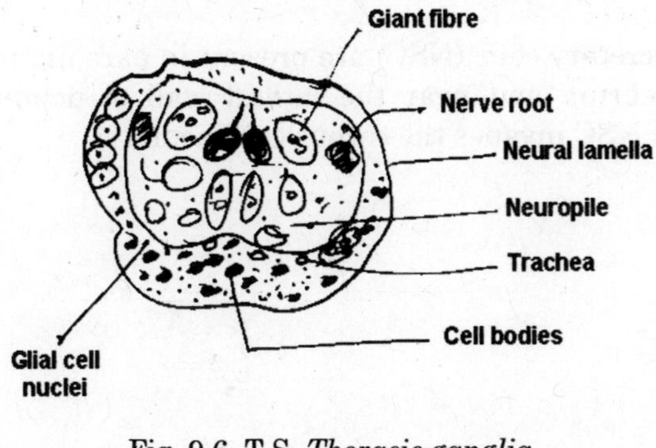

Fig. 9.6. T.S. *Thoracic ganglia*

7. T.S. Brain

1. Insect brain consist of three parts fused into a single mass
   a. *protocerebrum* : It reporesent united gagnlion of the optic segment and innervates the compound eye and ocelli.

b. *deutero cerebrum* : It is fused ganglion of antennary segment.

c. *tritocerebrum*: Formed by the ganglion of the 3rd or intercalary segment of the head.

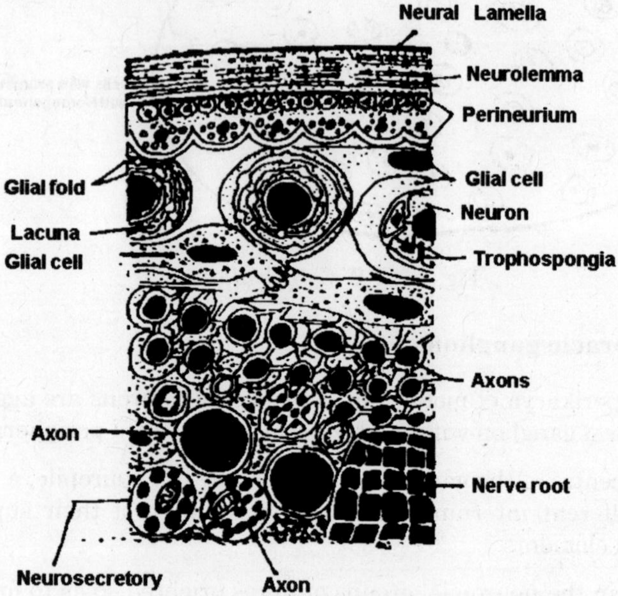

Fig. 9.7. Histology of the brain

2. Neurosecretary cells (NSC) are present in pars intercerebralis of protocerebrum and near the lateral wall of deuterocerebrum. Axons of NSC invades the corpus cardiacum.

CHAPTER - 10

MICROTOMY

Study of histology of a tissue involves *microtomy*. Cytological, histopathological and histochemical studies are also done with the help of this technique. It requires a set procedure which is followed step to step. Any fault in the steps involved can spoil the whole study. The procedure is being described in brief.

1. Fixation of material

The tissues are fixed in appropriate fixative. The specific tissues are dissected out with speed in physiological saline and fixed immediately. Bouin's fluid take minimum time (24 hrs) for proper fixation. Even the material can be left indefinitely in Bouin's fixative. Use 70% alcohol for insects. Cornoy fluid takes 2 to 4 hrs for fixation.

2. Washing

After the fixation, the tissue is washed properly so that the fixative goes out of the tissue. It is washed with distilled water or the jar can be kept under running tap water taking care so that the tissue may not be washed off. If the fixative is alcoholic then it is washed in alcohol. In case of cornoy's fluid they are washed in absolute alcohol (2-3 changes)

3. Dehydration and clearing

As usual it is passed through 30%, 50%, 70%, 90% and absolute alcohol. When fix in 70% then transfer the tissue in 90% for two hrs. Dehydration is necessary so that the wax can be deeply penetrated.

After dehydration it requires to be transparent which is done by clearing agent. Xylene is generally used. Do not leave the tissue in Xylene for more than the time it is required otherwise the tissue will be brittle. Cedar wood oil was also used but it requires more time. Methyl benzoate can also be used. The tissue are directly put into paraffin wax from the clearing agent. One or two changes of melted paraffin wax will remove the clearing agent completely from the tissue. It also facilitates proper penetration of the wax into the tissues.

4. Embedding and Block making

Embedding can be done using metal `L' blocks or paper boats. Paper boats can be easily made by using a wood piece. Paraffin wax is melted at 60-62°C temperature therefore keep the temperature of the oven at 60°C (Bees wax + paraffin, 1:3). After the wax is melted filter it so that all dirt can be removed otherwise the razor can be spoiled causing nicks to it.

Pour the melted wax in `L' moulds or paper boats and transfer the tissue into it. Orient it properly taking every care and mark the orientation of the tissue by a arrow mark with pen or pencil (in case of paper boats). It requires care and expertise. After the solidification of the wax in `L' moulds or paper boats, remove it and put it into the tray of water. It solidify properly now. Blocks are now trimmed properly. The trimmed blocks are mounted on the block holder of the Microtome.

5. Section Cutting

Set the microtome at the desired thickness at which the sections are to be cut (0.5 or 0.6 mm). At 0.7 mm thick sections will be there. Now start rotating the wheel. As soon as the block mounted on the holder makes a contact with razor, the section of the wax will be cut. A continuous ribbon is in the view now. As soon as the ribbon is 1-2 inch long put a fine brush below it and move it away from the microtome. After a length of about 6-7 inch cut the ribbon and place it on the clean slide. Take another ribbon likewise. For proper adherence of sections, it is required that a small drop of Mayer's albunen is rubbed on the slide. The sections are properly affixed now. Put the slide on the hot plate so that the sections are properly spread out along with slight melting of wax. Label the slide by using diamond point pencil.

6. Staining

All the wax is removed from the sections first. Put the slide in a couplin jar filled with xylene (make two changes) then to absolute alcohol for 5-6 minutes. It is dehydrated descendingly upto 30% alcohol and then to distilled water (two changes of 5 minutes each). Now use the aquous stain haematoxyline. Keep the slide under-running tap water to remove the excess stain and proceed the slide for dehydration in ascending series i.e. 30%, 50%, 70% alcohol. Now stain it with eosin. Bring it back to 70% then 90% alcohol, absolute alcohol and xylene (5-10 minutes). Mount the slide in D.P.X. Keep it for drying at room temperature or in oven at 60°C for few hours.

SUMMARY OF THE PROCEDURE

Tissue fixed (70% alcohol)

↓

90% alcohol (2 hrs)

↓

absolute alcohol (1/2 hr)

↓

Xylene (20 minutes)

↓

Bess wax + Paraffin wax . . . . . (Use less bees wax in summers)
(1 : 3)

↓

Wax + xylene (15 minutes)
50 : 50

↓

Wax pure (45 mintues)

↓

Block preparation

↓

Trimning of blocks

↓

Section cutting

↓

Ribbon of sections

↓

Spreading of sections

↓

Drying (overnight)

Xylene (1/2 hr two changes of 20+10 minutes)

↓

absolute alcohol (5-6 minutes)

↓

90% alcohol (5 minutes)

↓

70% alcohol (5 minutes)

↓

50% alcohol (5 minutes)

↓

30% alcohol (5 minutes)

↓

Distilled water

↓

Haematoxylin staining

↓

Running tap water (1/2 hr.)

↓

30% alcohol (5 minutes)

↓

50% alcohol (5 minutes)

↓

70% alcohol (5 minutes)

↓

Eosin staining (2-3 minutes)

↓

70% alcohol (3 minutes)

↓

90% alcohol (3 minutes)

↓

absolute alcohol (3 minutes)

↓

Xylene (5 minutes)

↓

DPX mounting

↓

Drying

BRIEF DESCRIPTION OF A MICROTOME

Rotary microtomes are generally used in all laboratories. Rocking type microtomes are also used. It has a very heavy base.

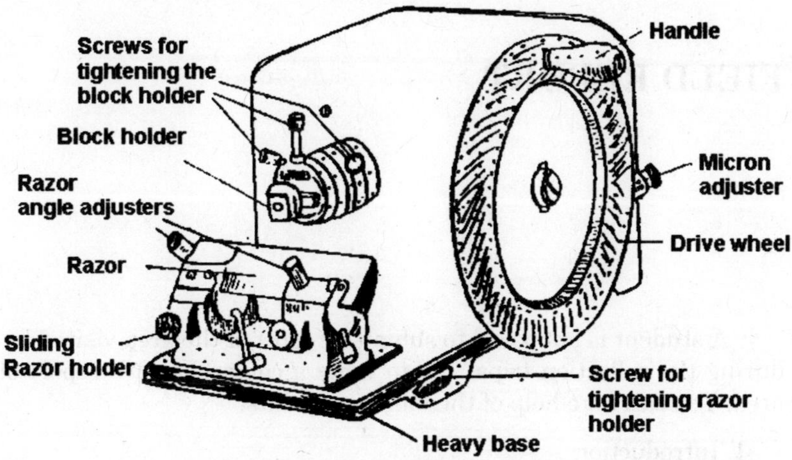

Fig. 10.1. Rotary microtome

Quite a few new models of Rotary microtomes are available. A circular drive wheel is rotated which cause the movement of axle forward. As the front part which holds the block holder moves up and down, the central axle moves forward simultaneouly by the same distance. The razor is mounted on a pair of hands with a groove and adjusting screws. The razor can be moved forward and backward along the grooves so that the razor can also be brought nearer to the block if required. On a desired position it is locked with the help of levers.

CHAPTER - 11

FIELD REPORT

A student is supposed to submit a report of the area visited by him during the collection trips. Try to write a comprehensive report of the area. You can take help of the following points.

1. Introduction

2. Topography of the area visited, vegetation types etc.

3. Geographical coordinates and nearest city, or district or highway.

4. (a) If the area is a lake or aquatic body or pond or river then

(i) Catchment area and historical background

(ii) Capacity

(iii) Present water level

(iv) Present uses of the water body

(v) Threats with special reference to Insects. (faunal diversity)

(vi) Inter specific relations if any

(vii) Agricultural activity in the area and crops grown in both Rabi and Kharif seasons.

(viii) Use of pesticides, (prepare a list of pesticides used in the area by surveying pesticide dealers and user farmers).

(ix) Principal pests identified in the area and their seasonal abundance.

(x) Notes on their natural enemies (if any)

(xi) Map of the area and photographs.

(xii) Suggestions or recommendations.

If the area is a forest then

(i) Total forest area

(ii) Vegetation of the forest and (flora)

(iii) Map of the area

(iv) Legal status of the forest. Past and Present use. (whether protected or not or private property).

(v) Fauna with special reference to Insects (faunal diversity)

(vi) Host plants in case of Butterflies, their population estimates, first hand information on roosting and natural enemies

(vi) Threats identified by you (as wood cutting, destruction, illegal activity, if any).

(vii) Photographs etc.

(viii) Suggestions and recommendations.

REFERENCES AND SUGGESTED READINGS

Bickley, W. 1942. On the stomodeal nervous system of insects. *Annals. Ent. Soc. Amer.*, 35 : 343-354.

Butt, F.H. 1943. Comparative study of mouthparts of representative Hemiptera-Homoptera. Cornell. U. Ag. Exp. Sta., *Memoir*. 254: 1-20.

Hood, C.W. 1937. The anatomy of digestive system of *Oncopeltus fasciatus* Dall. *Ohio. J. Sci.* xxxvii : 151-160.

Ross, H.H. 1930. Notes on digestive and reproductive system of the German Cockroach. *Ill. State Acad. Sci.,* 22 : 206-216.

Trigunayat, M.M. and Singh, N.P. 1998. Checklist of Common Butterflies of Jaipur region. *Geobios New Report* 12 (3&4): 167-171.

Trigunayat, M.M. & Singh, N.P. 1998. Effect of Penfluron on eggs of *Lasioderma serricorne.* Abst. Indian Sci. Congr. Hyderabad.

Trigunayat, M.M. 1999. Status of Butterflies of Keoladeo National Park, Bharatpur. *J. Exp. Zool. India,* 2 (1) : 55-57.

Trigunayat, M.M. and Singh, N.P. 1997. Laboratory studies- on the development of *Lasioderma serricorne* Fabr. *Nat. Symp.* on Developmental Biology, Orissa.

Trigunayat, M.M. 1997. Studies on the effects of Penfluron and Fenvale rate on Insect pests. Ph.D. Thesis. Dept. of Zoology, University of Rajasthan, Jaipur.

Books

Alven M. Erorrs: General Textbook of Entomology Packard - Treatise on injurous and Beneficial insect to crops.

Atwal, A.S. (1988) Agricultural pests of India and South east Asia, New Delhi.

Avasthi, V.B. (1997) Introduction to General and Applied Entomology, Jodhpur.

Chapman, R.L. (1971) Insect Structure and function, London.

Duncan, Martin: The transformation of insects.

Elzinga, R.J. (1978) Fundamentals of Entomology, New Delhi.

Essig, E.O. (1942) College Entomology. McMillan Co.

Frost, S.W. (1959) Insect Life and Insect Natural history, New York.

Fox, R.M. and Fox, J.W. (1964) Introduction to Comparative Entomology, New York.

Hati, A.K. (1979) Medical Entomology, Calcutta.

Kalyanam, N.P. Common Insect of India, Asia Publishing House.

Leffroy, H.M. (1984) Indian Insect Life, New Delhi.

Little, V.A. (1973) Pure and Applied Entomology, New Delhi.

Mani, M.S. (1982) Text Book of Entomology, New Delhi.

Mani, M.S. Insects, NBT, New Delhi.

Ross, H.H. (1965). Text Book of Entomology, John Wiley & Sons. Newyork.

Saxena, S.C. (1993) Biology of Insects, New Delhi.

Singh, N.P. (1999) Refresher course in Entomology ASC, Univ. of Raj. Jaipur.

Sen Gupta (1982) Insect Collection & Preservation, ZSI, Calcutta.

Srivastava, K.P. (1993) Text Book of Applied Entomology, Vol. II, New Delhi.

Snodgrass, R. (1971) Principles of Insect morphology, New Delhi.

Sutherland, W.J. (1996). Ecological Census techniques. Cambridge Univ. Press.

Tembhre, D.B. (1997) A Text book of Entomology, HPH, Bombay.

Tonapi, G.T. (1994) Experimental Entomology - An aid to field and laboratory, New Delhi.

Trigunayat, M.M. (2001). Practical Zoology, S.P.H., Jaipur.

Trigunayat, M.M. (2004): A Manual of Practical Zoology. SPH, Jaipur.

Richard, D.W. & Davis, R.G. - Imms Introduction to Entomology, New Delhi.

Wigglesworth, V.B. — Principles of Insect Physiology, London.

APPENDIX — I

1. Insect's Ringer's solution

Nacl	0.65 gm
Sodium carbonate	0.02 gm
Calcium chloride	0.03 gm
Potassium chloride	0.025 gm
Distilled water	100 ml

Dissolve the contents.

2. Physiological saline

Dissolve 0.6 gm Nacl in 100 ml. of distilled water.

3. Mounting media

(i) Berlese medium

Chloral hydrate	160 gm
Gum arabic	15 gm
Glucose syrup	10 gm
Acetic acid	5 ml
Distilled water	20 ml

(ii) de Faure's medium

Chloral hydrate	50 gm
Gum arabic	30 gm

Glycerine	20 ml
Distilled water	50 ml

(iii) Hoyer's medium

Chloral hydrate	200 gm
Gum arabic	30 gm
Glycerine	20 ml
Distilled water	50 ml

4. Killing and Preservative fluids

(i) Kerosene, alcohol, acetic acid and Dioxine (KAAD)

Kerosene	5 ml
95% ethanol	50 ml
Glacial acetic acid	20 ml
Dioxine	10 ml

(Material should be transferred to 70-95% alcohol with in 1-4 hours)

(ii) Barber's fluid

95% ethanol	265 ml
Ethylacetate	95 ml
Distilled water	245 ml
Benzol	35 ml

(Material is transferred to 70% alcohol within 24 hrs)

5. Stains and other reagents

(i) Eosin (aquous)

Eosin powder	1 gm
Distilled water	100 ml

(ii) Eosin (alcoholic)

Eosin powder	1 gm
70% alcohol	100 ml

(iii) Acetocarmine

2 gm carmine powder + 100 ml 45% boiled acetic acid

Boil for 10 minutes and filter

(iv) Acetoorcein

2 gm Orcein powder + 100 ml 45% acetic acid boiled.

Boil for 10 minutes and filter

(v) Acid Fuchsin

10% Hcl	25 ml
Acid fuchsin	0.5 gm
Distilled water	300 ml

Mix the ingradients and use as stain

(vi) Lactophenol

60 ml acetic acid + 60 ml carbolic acid.

Heat in water bath till the bubbles are driven out.

(vii) Polyvinyl alcohol

Dissolve 5 gm polyvinyl alcohol in 30 ml distilled water.

(viii) 10% KOH

10 gm KOH in 100 ml distilled water.

(ix) Haematoxylin (Ehrlich)

Dissolve haematoxylin in absolute alcohol

Haematoxylin	2 gm
Absolute alcohol	100 ml

add the following in sequence

glycerine	100 ml
distilled water	100 ml
glacial acetic acid	10 ml

potassium alum in excess

(x) Acid alcohol

Alcohol 70%	100 ml
Conc. HCl	2.5 ml

(xi) Acid water

Distilled water	100 ml
Conc. HCl	2.5 ml

(xii) Schiff's reagent

Basic fuchsin	2 gm
Potassium metabisulphite ($K_2S_2O_5$)	3.8 gm
0.15 NHCl	200 ml

(xiii) Sulphite wash

10% sodium metabisulphite	30 ml
1 NHCl	30 ml
Distilled water	500 ml

Add metabisulphite into distilled water and then add acid.

(xiv) Carnoy's fixative

Absolute alchohol (ethyl)	60 ml
Glacial acetic acid	10 ml
Chloroform	30 ml

(Fix the tissue at least 15-60 minutes)

(xv) Fast green

Fast green 0.01% in 95% alcohol or 1 gm in 95% alcohol

(xvi) Bouin's fixative

Picric acid	75 ml
40% HCHO	25 ml
Glacial acetic acid	5 ml

(xvii) FAA

40% HCHO	10 ml
70% Ethyl alcohol	85 ml
Glacial acetic acid	5 ml

(xviii) Normal saline

Nacl	0.78 gm
Distilled water	100 ml

(xix) Leishman stain

Leishman stain	0.15 gm
Absolute alcohol	100 ml

(xx) Congo red

Congo red	0.5 gm
Distilled water	100 ml

(xxi) Crystal violet

Crystal violet	2 gm
Ethyl aclohol (90%)	20 ml
Distilled water	180 ml

(xxii) Giesma stain

Giesma powder	0.5 gm
Glycerine	33 ml

Absolute alcohol (methyl) 33 ml

Keep the mixture over night for incubation at 60°C & filter.

Note: Giesma is a mixture of methylene blue, Azure A & B Staining solution is made by diluting stock solution.

(2.5 ml stock solution + 1 ml methanol + 50 ml Giesma water (pH 6.9).

(xxiii) Methylene Blue

Methylene blue	1.5 gm
Ethyl alcohol (90%)	100 ml

(xxiv) Zenker's solution

$K_2Cr_2O_7$	2.5 gm
$HgCl_2$ (Mercuric chloride)	15 gm
Distilled water (DW)	100 ml
Sodium sulphate	1 gm
Glacial acetic acid	5 ml

(xxv) Karl's fluid

Ethyl alcohol 90%	170 ml
40% HCHO	60 ml
DW	280 ml
Glacial acetic acid	20 ml

(Add acetic acid just before using)

(xxvi) Neutral red

1 part neutral red + 3000 parts ethyl alcohol

(xxvii) Iodine (Lugol's)

KI	10 gm
DW	100 ml Dissove and add iodine
Iodine	5 gm

(It is good for cilia, Flagella & Nucleus)

Aquous Iodine

KI	2 gm
DW	100 ml
Iodine Crystal	1 gm

(xxviii) Picro indigo carmine

Picric acid (saturated solution in 90% alochol) 1 part

Indigo solution in 70% alcohol 2 parts

(It is used for chitinous material)

(xxix) Kleinberg's fixative

Picric acid + 2% aquous solution of H_2SO_4

(xxx) Janus green (Diethyl safranin azodimethyl aniline)

Prepare 1% aquous solution.

6. Artificial diet for Drosophila

(i) Agar agar or gelatin 10 gm

Take 300 ml of water and warm it, add agar agar powder till it dissolves.

(ii) non crystal sugar	65 gm
Corn flour	75 gm
yeast powder	10 gm

Add all these to 450 ml of water, stir it continuously to avoid lumping and mix both ingradients (a) and (b). Boil and add 2 ml of propionic acid when thin paste is formed, pour it into the culture bottles while still hot. Allow to cool at room temperature. Wipe out the moisture from the inner walls of the bottle.

7. Artificial diet for *Helicoverpa* (= Heliothis)

Agar	10.20 gm
Yeast	5 gm

Ascrobic acid	2.6 gm
Methyl para hydroxy benzoate (MPH)	1.60 gm
Sorbic acid	0.80 gm
Multivitaplex	0.60 gm (Dumex company)
Streptomycin sulphate	0.20 gm
Formaldehyde 10%	1.60 ml
Distilled water	400 + 224 ml

Food for adult - Sugar solution 20% or Honey solution 10% + 1% Protinex.

APPENDIX - II

GRAPHIC REPRESENTATION OF LD_{50} BY PROBIT ANALYSIS

Calculation of Probit values

$$y = 5 + \frac{1}{\sigma}(x - \mu)$$

y = probit

σ = standard deviation of log dose

x = log dose

μ = mean of log dose

Standard deviation can be calculated by the following formula

$$\sigma = \sqrt{\frac{\left[\Sigma x^2 - \frac{(\Sigma x)^2}{n}\right]}{n - 1}}$$

Dose	2	4	8	16	32
Log dose	0.3010	0.6020	0.9030	1.2041	1.5051

x = [0.30, 0.60, 0.90, 1.20, 1.50] = Σ x 4.5

Σx^2 = [0.09, 0.36, 0.81 1.44 2.25] = 4.95

$$\Sigma x = 4.5, \quad \frac{[\Sigma x]^2}{n} = \frac{(4.5)^2}{5} = \frac{20.25}{5} = 4.05$$

$$\sigma = \sqrt{\frac{4.95 - 4.05}{5 - 1}}$$

$$\sigma = \sqrt{\frac{0.90}{4}} \qquad \sigma = \sqrt{0.225} \qquad \sigma = 0.47$$

$$\mu = \frac{\Sigma x}{n} \quad \mu = \frac{4.5}{5} = 0.90$$

Now put the value in the formula

$$y = 5 + \frac{1}{\sigma}(x - \mu)$$

$$y = 5 + \frac{1}{0.47}(0.30 - 0.90)$$

$$y = 5 - \frac{1 \times 0.60}{0.47}$$

$$y = 5 - 1.2$$

$y = 3.8$ Probit value at log dose 0.30

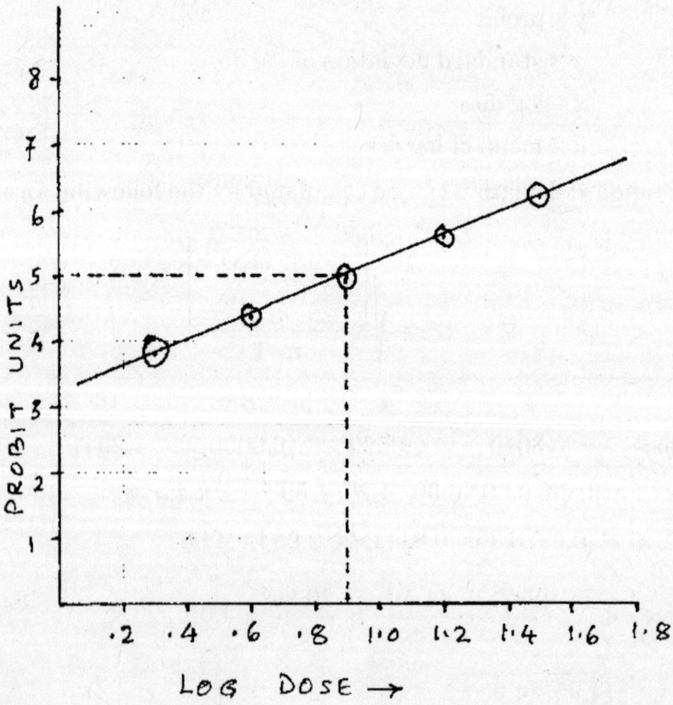

Similarly at dose 4 log dose 0.60

$$y = 5 + \frac{1}{0.47}(0.60 - 0.90)$$

$$y = 5 - \frac{0.30}{0.47}$$

$$y = 5 - 0.63$$

$$y = 4.4$$

Similarly at dose 8 log dose 0.90

$$y = 5.0$$

at dose 16 log dose 1.20

$$y = 5.6$$

at dose 32 log dose 1.50

$$y = 6.2$$

The probit values thus calculated are 3.8, 4.4, 5.0, 5.6 & 6.2 respectively.

Now plot the graph log dose versus probit values.

APPENDIX - III

Identification of insects with the help of taxonomic keys

Order - COLEOPTERA

KEY TO SUBORDERS

1. Head not prolonged into a beak; gular sutures double, at least anteriorly and posteriorly; prosternal sutures distinct.... 2

   Head usually prolonged into a beak; gular sutures fused or lacking; prosternal sutures lacking. Vegetable feeders. .... *Rhynchophora*

2. Outer lobe or galea of maxillae palpiform; first visible abdominal sternite divided by the hind coxal cavities; hind wings with one or two cross veins near the middle; pleural sutures of prothorax present; antennae filiform or nearly so; tarsi five segmented; larvae thysanuriform, the tarsi with one or two claws. Largely predacious or carnivorous.................. *Adephaga*

   Outer lobe or galea of maxillae not palpiform; first visible abdominal sternite not usually interrupted by the hind coxal cavities; hind wings without cross veins; pleural sutures of prothorax absent; antennae and tarsi variable; larvae variable, the tarsi and claws fused. Animal and vegetable feeders.. *Polyphaga*

A. Suborder ADEPHAGA

KEY TO SUPERFAMILIES

1. Abdomen with six or more visible sternites; antennae filiform 2

Abdomen with four visible sternites; antennae greatly thickened apically, clavate or broadly laminate. Myrmecophilous .. ..*Paussoidea*

2. Eyes entire, not divided; antennae elongate, slender. Predacious..... .... *Caraboidea*

Eyes divided, appearing as two pairs; antennae very short, robust, irregular.. *Gyrinoidea*

I. Superfamily CARABOIDEA

KEY TO MORE IMPORTANT FAMILIES

1. Metasternum with a distinct transverse suture demarking a triangular antecoxal sterite; posterior legs without dense fringes of long hair.. 2

Metasternum without a transverse suture or antecoxal sclerite; posterior legs with dense fringes of long hair. (Predacious Diving Beetles)... ....*Dytiscide*

2. Transverse suture in front of hind coxae extending across the metasternum which is continued behind in a triangular process between the hind coxae... 3

Transverse suture in front of hind coxae very short, reaching only across the central portion; metasternum not prolonged between the hind coxae; front coxae globular; tibiae and tarsi without swimming hairs; small dull aquatic or semiaquatic beetles. Rare ....*Amphizoidae*

3. Hind coxae normal; antennae 11-segmented, with at least the six apical segments pubescent; terrestrial... 4

Hind coxae fixed and extended into two broad plates covering the femora and the base of the abdomen; antennae 10-segmented, without pubescence; hind legs without fringe of long hairs for swimming; aquatic. Small, oval, usually dark, water beetles... .... *Halipidae*

4. Clypeus extending on each side beyond the base of the antennae; antennae inserted on the front above the base of the mandibles; inner lobe of lacinia terminating in a movable hook; eyes large; head vertical, wider than thorax. Usually medium sized, slender, brightly colored or metallic blue and green beetles with yellow marks. They occur in damp places and along waterways. Very alert and active. (Tiger beetles).. *Cicindelidae*

Clypeus not extending beyond the base of the antennae; antennae arising from the sides of the head between the base of mandibles and the eyes; inner lobe or lacinia of maxillae not terminating in a movable hook; eyes normal; head horizontal or slightly inclined, usually narower than thorax. Small to large, usually black, but may be brightly colored and metallic reddish, blue and green. Fast running. (Predacious Ground Beetles).... .... *Carabidae*

Superfamily CANTHAROIDEA

KEY TO IMPORTANT FAMILIES

1. Abdomen with seven or eight visible sternites.... 2

Abdomen with six visible sternites..... 4

2. Middle coxae distinct; epipleura wanting; elytra usually with reticulated surface; no phosphorescent organs. Largely tropical beetles. (Net winged Beetles).. *Lycidae*

Middle coxae in contact; epipleura distinct; elytra not reticulate.................. 3

3. Episterna of metathorax sinuate on inner side; head nearly or quite covered by thorax; epipleura usually wide at base of elytra; phosphorescent organs usually present. (Fireflies.).... ....*Lampyridae*

Episterna of metathorax not sinuate on inner side; head fully exposed; epipleura narrow at base; anterior coxae with distinct trochantins; elytra soft; no phosphorescent organs. Medium sized predaciou beetles. (Leather winged Beetles.)..*Cantharidae*

4. Hind coxae prominent, at least internally.. 5

Hind coxae flat, not prominent.... 6

5. Body with extensible vesicles... ....*Malachiidae*

Body without extensible vesicles... .... *Melyridae*

6. Fourth segment of tarsus small; pronotum separated from propleura ....*Corynetidae*

Fourth segment of tarsus normal; pronotum not separated from propleura.... .... *Cleridae*

Superfamily ELATEROIDEA

KEY TO IMPORTANT FAMILIES

1. First two abdominal sternites fused or immovable; tarsi with membranous lobes beneath; thorax and abdomen firmly united; antennae serrate, 11-segmented. Small to large, often beautiful metallic

and iridescent beetles. (Metallic Wood-Boring Beetles.) .... *Buprestidae*

All the abdominal segments free and movable.. 2

2. Prothorax firmly attached to the mesothorax, not movable, front coxal cavities closed behind by mesosternum; prosternal process prolonged but not movable. Minute black or brown beetles. (Pseudo Click Beetles.).. ....*Throscide*

Prothorax loosely joined to the mesothorax and freely movable, the hind angles each prolonged into a tooth; prosternal proces loosely received in a socket in the mesosternum; front coxal cavities entirely within the prosternum.. 3

3. Prosternum lobed in front; labium visible; prothoracic process movable and when thrust into the socket on the mesosternum and forcibley snapped out causes the beetle to jump and click. Small to large, mostly brown and black beetles. (Click Beetles.)... .... *Elateridae*

Prosternum not lobed; labium concealed; antennae inserted on front, somewhat distant from eyes; not able to jump and click. Small and rare beetles.... ....*Eucnemidae*

COLEOPTERA

Superfamily DRYOPOIDEA

Family DRYOPIDAE (= PARNIDAE)

Small, pubescent and hairy subaquatic and aquatic beetles. The adults carry air on the bodies and crawl about rather than swim. The larvae of *Dryops* Olivier live under stones. They are cosmopolitan. The members of related family PSEPHENIDAE are aquatic and have remarkable disk-like larvae that attach themselves to stones in swift-flowing streams.

Superfamily DASCILLOIDEA

KEY TO IMPORTANT FAMILIES

1. Front coxae transverse, subcylindrical 2

Front coxae globular or conical.... 3

2. Legs stout, retractile; front coxae open behind; hind coxae extending to the margin of the body; tibiae dilated, usually with a groove near the apical end for the reception of the tarsi; tibial spurs distinct; last tarsal segment small and quadrate. Small to medium convex beetles. (Pill Beetles.).... .... *Byrrhidae*

Legs normal, tibiae slender, spurs reduced or absent; front trochantins present; antennae serrate, 11-segmented, inserted below a slight ridge in front of the eyes; claws simple or pectinate. Small to medium terrestrial and aquatic beetles. (Soft-bodied Plant Beetles.) .... *Dascillidae*

3. Front coxal cavities closed behind; plate of hind coxae weak. Small oval, pubescent beetles found in flowers .... *Byturidae*

Front coxal cavities open behind; plate of hind coxae distinct. Small convex scaly beetles usually feeding on dead or dry animal matter. (Skin or Larder Beetles.)... .... *Dermestidae*

Superfamily CUCUJOIDEA

KEY TO IMPORTANT FAMILIES

1. Tarsi three-segmented.... 2

Tarsi four or five-segmented.... 3

2. Tarsal claws simple; mesepimeron quadrangular; first abdominal sternite without coxal lines. Oblong or suboval, brightly colored beetles, mostly living in fungi...*Endomychidae*

Tarsal claws dilated or toothed at base; mesepimeron traingular; first abdominal sternite with arcuate coxal lines. Small oval predacious beetles. (Ladybird Beetles.)..*Coccinellidae*

3. Tarsi five segmented, rarely heteromerous.. 4

Tarsi four segmented; first three or four abdominal sternites fused; antennae 10- to 11-or rarely eight segmented terminating in a one or two segmented club; front and middle coxae small and globular, hind coxae transverse. Small elongate or cylindrical bark or ground beetles. (Cylindrical Barkbeetles)...*Colydidae*

4. Front coxae transverse... 5

Front coxae not transverse 6

5. Tarsi more or less dilated, first segment not shortened; fourth segment very small; elytra truncate, not usually extending to the tip of the abdomen. (Sap-feeding Beetles).... ....*Nitidulidae* (in part)

Tarsi slender, the first segment very small; elytra never truncate, always covering abdomen. Small to medium, usually flattered beetles living under bark or inhabiting granaries.
(Grain and Bark gnawing Beetles)...*Ostomatidae*

6. Front coxae globose, closed behind; tarsi dilated and hairy, fourth segment very small; antennae 11-segmented, the last three or four segments forming a club; elytra entire; head immersed into the thorax almost to the eyes. Small to medium sized red and black beetles. (Pleasing Fungus Beetles.)...*Erotylidae*

Front coxae oval... 7

7. Middle coxal cavities not closed on the outer side by the union of the meso-and metasterna; tibiae with two spurs; antennae 11 segmented; elytra covering abdomen. Small, very flat black, brown, or reddish beetles often living under bark. (Flat Barkbeetles).. ....*Cucujidae*

Middle coxal cavities closed by the sterna; posternum prolonged; front coxal cavities open behind; antennae 11 segmented with loose club. Very small oval or elongate, flat, pubescent, punctate, and often yellow or brown beetles. (Silken Fungus Beetles.).. *Cryptophagidae*

Superfamily MORDELLOIDEA

KEY TO IMPORTANT FAMILIES

1. Head strongly constricted behind the eyes; tarsal claws sometimes pectinate.... 2

Head gradually narrowed behind the eyes; tarsal claws simple or cleft; middle coxae prominent, contiguous. Slender, soft bodied beetles .... *Cedemeridae*

2. Sides of prothorax with sharp lateral margins, the base as wide as the elytra; antennae filiform, 11-segmented; head joined vertically against the thorax; hind coxae flat; elytra soft; tip of abdomen prolonged into a point. (Tumbling Flower Beetles.) ....*Mordellidae*

Sides of prothorax rounded and without sharp lateral margins.................... 3

3. Hind coxae large and prominent; tarsal claws cleft or toothed, head deflexed, front vertical; antennae 11-segmented, inserted at the sides in

front of the eyes; elytra soft and entirely or only partly covering abdomen. Medium to large soft-bodied beetles. (Blister Beetles.) .... *Meloidae*

Superfamily PTINOIDEA

KEY TO FAMILIES

1. First tarsal segment longer than the second... 2

First tarsal segment shorter than the second.. 3

2. Antennae inserted on the front of the head close togethe at the base; thorax without lateral margin; head retractile, front and middle coxae cylindrical, hind coxae transverse. Small to medium-sized, elongated, robust, brown or blackish beetles. (Ptinid Beetles, Spider Beetles). ....*Ptinidae*

Antennae inserted on the sides of the head in front of the eyes and usually distant at the base; thorax usually with lateral margins. Small brown or black, elongated and depressed beetles. (Death Watch Beetles.)..*Anobiidae*

3. First abdominal segment longer than the others; antennae terminating in two saw-toth egments; head prominent, free. Elongated slender black or brown beetles. (Powder Post Beetles.) ... .*Lyctidae*

First abdominal segment equal in length to the others; antennae terminating in three or four saw-toothed segments, head deflexed and often hidden by prothorax. Usually robust or cylindrical brown or black beetles. (Branch and Twig Borers.) .... *Bostrichidae*

Superfamily SCARABAEOIDEA

KEY TO IMPORTANT FAMILIES

1. Lamellate segments of antennal club comb-like or pectinate and not capable of being closely coadapted.. 2

Lamellate segments of antennal club thin and leaf-like and capable of being closely coadapted; antennae not elbowed; abdomen with five or six visible rigid or free sternites; elytra rough, smooth, hairy, or scaly. Small to large robust dull or brightly colored, often metallic beetles. (June Beetles, Lamellicorn Beetles, Cockchafers, etc.) 3

2. Antennae not elbowed but usually curved so as to appear elbowed; mentum deeply emarginate, the ligula large, corneous, and filling the emargination; front of head with a short bent horn pointing forwards; body pedunculate; elytra striate and glabrous. Large oblong shining brown or black beetles. (Horned Beetles.) ....*Passalidae*

Antennae usually elbowed, rarely straight; mentum entire and usually completely covering the maxillae; mandible prominent and strongly developed in the males. Large smooth, rough, striate or punctate black and brown beetles. (Stag Beetles.)................................... *Lucanidae*

3. Abdomen with six tergites; side pieces of metasternum attaining coxae....4

Abdomen with five tergites; side pieces of metasternum not attaining coxae; body roughly sculptured, elytra frequently tuberculate, (Skin Beetles.)... ....*Trogidae*

4. Mentum and ligula fued; abdominal spiracles all on membrane between tergites and sternites.. 5

Mentum and ligula fused; abdominal spiracles at least in part upon sternites... 10

5. Hind tibiae with a single terminal spur; scutellum small or invisible; middle legs widely separated. (Tumble-bugs, Scarabs, etc.) .... *Scarabaediae*

Hind tibiae with two terminal spurs; scutellum distinct; middle legs approximate... 6

6. Antennae 11-segmented.... 7

Antennae not more than 10-segmented.. 8

7. Antennal club three-lamellate.... .... *Geotrupidae*

Antennal club four-to seven lamellate..*Pleocomidae*

8. Antennae nine-segmented...9

Antennae 10-segmented; body very hairy; metepimera large; antennal club glabrous. ....*Glaphyridae*

9. Mandibles concealed beneath clypeus... ....*Aphodiidae*

Mandibles exposed; not covered by clypeus... ....*Aegialiidae*

10. Abdominal spiracles in two only slightly posteriorly divergent rows, each row forming a single straight line........................ .. *Melolonthidae*

Abdominal spiracles in two strongly poteriorly divergent rows, each row forming tow lines 11

11. Claws of posterior tarsi subequal in length... 12

Claws of posterior tarsi unequal in length, movable; pygidium exposed .... *Rutelidae*

12. Mandibles expanded, visible from above.... .... *Dynatidae*

Mandibles not expanded, not visible from above........ *Cetoniidae*

Superfamily CERAMBYCOIDEA

KEY TO FAMILIES

1. Mentum pedunculate; front prolonged into a broad quadrate beak; antennae short, serrate or pectinate, inserted in front of the eyes; prosternum very short; hind femora usually thickened; body short, more or less oval, tip of the abdomen exposed. Small robust seed-infesting beetles. (Pea and Bean Weevils.).... .... *Lariidae*

Mentum not pedunculate 2

2. Body short and somewhat oval; antennae short or not greatly lengthened and not inserted on frontal prominences; front small, oblique or inflexed; thorax frequently margined, tibial spurs usually absent. Minute to mall oval dull or brightly colored, smooth, hairy or scaly plant-eating beetles. (Leaf Beetles.)........ .*Chrysomelidae*

Body elongate; antennae usually long and often greatly lengthened and frequently inserted on frontal prominences; front often large, vertical, and quadrate; thorax rarely margined; tibial spurs distinct. Medium to large slender or cylindrical dull or brightly colored, smooth, punctate, hairy or scaly wood boring beetles. The adults of some frequent flowers. (Long horned Beetles)........ *Cerambycidae*

Suborder RHYNCHOPHORA
Superfamily BRENTOIDEA
Superfamily CURCULIONOIDEA

KEY TO IMPORTANT FAMILIES

1. Beak abortive or extremely short and broad; antennae short with large club; tibiae often with several external teeth; small oval or cylindrical brown or black beetles.... 2

Beak well developed antennae without or with but a small club; tibiae without external teeth... 3

2. First tarsal joint as long as others together; head wider than the prothorax; eyes rounded.... .... *Platypodidae*

First tarsal joint shorter than the combined length of the others; head narrower than the prothorax; eyes oval, emarginate or divided. (Barkbeetles.).... .... *Scolytidae*

3. Antennae usually elbowed, with the basal joint greatly elongated; palpi small, short, rigid, and nearly always concealed within the mouth; beak often curved downwards. (Weevils.) (Including RHINOMACERIDAE, RHYNCHITIDAE, ATTELABIDAE, BRYSOPIDAE, OTIORHYNCHIDAE, CALANDRIDAE).... .... *Curculionidae*

Antennae not elbowed; palpi usually exposed.... 4

4. Prothorax elongate, elytra covering the pygidium, first two ventral abdominal segments fused; beak short and broad.... .... *Anthribidae*

Prothorax not elongate, usually trapezoidal; pygidium expoed; ventral abdominal segments free; beak very long in the female, sometimes entirely absent in the male.... .... *Brentidae*

Order - HEMIPTERA

CLASSIFICATION

Suborder HETEROPTERA

SERIES I. GYMNOCERATA
(Terrestrial and Subaquatic Forms)

Section GEOCORISAE

1. Family *PLATASPIDIDAE*
2. Family *PODOPIDAE (GRAPHOSOMATIDAE*)
3. Family *CYDNIDAE*
4. Family *SCUTELLERIDAE*
5. Family *PENTATOMIDAE*
6. Family *COREIDAE*
7. Family *CORIZIDAE*
8. Family *ALYDIDAE*

9. Family *ARADIDAE*
10. Family *DYSODIDAE*
11. Family *TERMITAPHIDIDAE*
12. Family *NEIDIDAE (BERYTIDAE)*
13. Family *LYGAEIDAE*
14. Family *PYRRHOCORIDAE*
15. Family *TINGIDAE*
16. Family *PIESMIDAE*
17. Family *THAUMASTOTHERIIDAE (THAUMASTOCORIDAE)*
18. Family *ENICOCEPHALIDAE*
19. Family *PHYMATIDAE*
20. Family *REDUVIIDAE*
21. Family *PLOIARIIDAE*
22. Family *NABIDAE*
23. Family *Polyctenidae*
24. Family *Cimicidae*
25. Family *Anthocoridae*
26. Family *Microphysidae*
27. Family *Miridae (Capsidae)*
28. Family *Cryptostemmatidae (Ceratocombidae, Dipsocoridae)*
29. Family *Saldidae*

SECTION AMPHIBICORISAE

30. Family *Hebridae*
31. Family *Mesoveliidae*
32. Family *Hydrometridae*
33. Family *Veliidae (Microveliidae)*

34. Family *Gerridae*

SERIES II. CRYPTOCERATA (HYDROCORISAE) (Aquatic Forms)

35. Family *Gelastocoridae (Nerthridae, Galgulidae, Mononychidae)*

36. Family *Ochteridae*

37. Family *Naucoridae*

38. Family *Nepidae*

39. Family *Belostomatidae (Belostomidae)*

40. Family *Notonectidae*

41. Family *Pleidae*

42. Family *Helotrephidae*

43. Family *Corixidae*

APPENDIX - III

KEY TO THE FAMILIES OF HETEROPTERA (Adapted from Usinger)

1. Antennae shorter than head, concealed in grooves beneath the eyes and hence invisible from above except in the OCHTERIDAE. Series CRYPTOCERATA.......... 2

Antennae longer than the head and plainly visible from above. Series GYMNOCERATA.......... 9

2. Ocelli absent.......... 3

Ocelli preent 8

3. Bae of head overlapping front margin of pronotum. Rostrum very short, scarcely distinguished from the broad apex of head. Front tarsi one-segmented, spatulate, and fringed with strong bristles. (Water Boatmen.).......... *Corixidae*

Base of head inserted in pronotum. Rostrum cylindrical or cone-shaped and distinctily segmented. Front tarsi not as above...4

4. Abdomen with a pair of short strap-like or long slender posterior appendages...5

Abdomen without such posterior appendages..6

5. Posterior abdominal appendages long and slender, forming a respiratory siphon. Hind legs cylindrical, formed for walking. Hind coxae globular. (Water Scorpions)..*Nepidae*

Posterior abdominal appendages short, flat, retractile, sometimes concealed beneath the tips of hemelytra. Hind legs flattened, fringed, fitted for swimming, the hing coxae hinged. (Giant Water Bugs) ..... *Belostomatidae*

6. Body broad, oval, and flattened. Front femora tremendously enlarged and grooved along inner edge where they oppose the prehensile, curbed tibiae. (Water Creepers.)... ....*Naucoridae*

Body strongly convex above, either elongate and large, over 5 mm., or more rounded and mall, less than 3 mm. Front femora not greatly enlarged....7

7. Form elongate. Size large 5mm. or more. Hind tibiae and tarsi densely clothed with swimming hairs. Abdomen with a ventral trough or concavity in which is a median carinate ridge. Antennae four segmented. (Back Swimmings).... ....*Notonectidae*

Form suboval. Size small, less than 3 mm. Hind tibiae and tarsi scarcely, if at all, clothed with swimming hairs. Abdomen neither deeply concave not carinate beneath..*Pleidae*

8. Antennae exposed. Eyes not strongly protuberant. Front legs similar to middle pair, fitted for running. (Velvety Shore Bugs) ....*Ochteridae*

Antennae concealed in grooves beneath the strongly protuberant eye. Front legs raptorial, the femora very broad and grooved along their inner edges nearet the curved tibiae. (Toad Bugs) .... *Gelastocoridae*

9. Claws of at least the front tarsi inserted before the apex, the tip of the lat tarsal segment cleft. Superfamily GERROIDEA 10

Claws all located at the tips of the tarsi, the tarsal apices entire...............11

10. Hind femora very long, greatly exceeding apex of abdomen. (Water Strider.)... *Gerridae*

Hind femora scarcely, if at all, surpaing tip of abdomen. (Smaller Water Striders).... .... *Veliidae*

11. Head as long as the entire thorax. (Marsh Treaders.)...... *Hydrometridae*

Head shorter, not exceeding the combined length of pronotum and scutellum... 12

12. Body clothed with short, velvety pile, at least ventrally. Both clavus and membrane membranous and devoid of veins. Small (less than 4 mm.) semiaquatic bugs with legs fitted for running or walking on the water...13

Body not clothed with fine, hydrofuge pile. Terrestrial. .... .14

13. Rostrum four segmented. Tarsi two segmented. Stout little bugs with short, subequal legs. (Velvet Water Bugs.) .... *Hebridae*

Rostrum three segmented. Tarsi three segmented, the basal segmented very small. Elongate, slender bugs with long slender legs for running on the surface of the water. (Water Treaders.)... .... *Mesoveliidae*

14. Head cylindrical or narrowed and stongly extuded or produced. Front legs moderately to distinctly enlarged, raptorial. Beak often short, stout, curved, and fitted for a predacious habit.. 15

Head usually broad and deeply inserted into the pronotum, and front leg usually slender. If the head is cylindrical or the front femora are enlarged and spined, then the membrane has only five simple longitudinal veins or the corium is deflected into a cuneus... 18

15. Head strongly constricted behind the eyes, the ocelli located on the posterior, subglobose lobe. Pronotum divided into three distinct lobes by transverse sutures. (Gnat Bugs.)... .... *Enicocephalidae*

Head less distinctly or not at all bilobed. Pronotum not divided into three lobes.. 16

16. Fourth antennal segment thickened. Front femora tremendously enlarged, subtriangular. (Ambush Bugs.)... ....*Phymatidae*

Fourth antennal segment not thickened, slender. Front femora scarcely to moderately incrassate... 17

17. Prosternum with a longitudinal, transversely striated, stridulatory groove between the coxae which receives the tip of the rostrum. Rostrum three segmented. (Assassin Bugs.)... ....*Reduviidae*

Prosternum without a longitudinal stridulatory groove. Rostrum usually four segmented. (Damsel Bugs.)........................*Nabidae*

18. Hemelytra with a distinct fracture subapically on the corium, producing a deflected cuneus.. .... .19

Hemelytra without a distinct cuneus... 22

19. Ocelli absent...(Leaf Bugs, Capsids.).. ....*Miridae*

Ocelli present.. 20

20. Tarsi two segmented. Body less than 1 1/2 mm. long. ...... *Microphysidae*

Tarsi three segmented.... 21

21. Rostrum four segmented. Apical two antennal segments without long, erect hairs. (Jumping Tree Bugs.).... ....*Isometopidae*

Rostrum three segmented. Apical two antennal segments usually beset with long hairs or bristles. (Flower Bugs.) ....*Anthocoridae*

22. Antennae four segmented.. 23

Antennae five segmented.... 34

23. Ocelli absent..... 24

Ocelli present... 28

24. Hemelytra reduced to short pads without trace of membrane. Flattened species with strongly asymmetrical male genital organ. Highly modified for an ectoparasitic, blood sucking existence...............25

Hemelytra complete or at least with some indication of membrane. Not modified as ectoparasites.. 26

25. Eyes absent. Ctenidia present, at least on the head and antennae. Hemelytra consisting of nonarticulating pronotal pads. Frong legs greatly enlarged. (Bat Bugs.).... ....*Polyctenidae*

Eyes present. Ctenidia absent. Hemelytra consisting of distinct but reduced, basally articulating wing pads. Front legs not greatly enlarged. (Bedbugs.)... ....*Cimicidae*

26. Tarsi three segmented. Body large, robust. (Red Bugs.) ... *Pyrrhocoridae*

Taris two segmented. Body quite strongly flattened.................................. 27

27. Hemelytra lace-like or closely reticulated. Pronotum prolonged backwards triangularly and nearly or quite covering scutellum. (Lace Bugs.).... .... *Tingidae*

Hemelytra not finely lace like in appearance. Pronotum not backwardly produced, the scutellum exposed. (Flat Bugs.) ... . *Aradidae*

28. Tarsi twi-segmented, the first segment often minute...........................29

Tarsi three segmented... 30

29. Rostrum four segmented. Membrane with a few distinct vein. Upper surface covered with fine areoles. (Ash-gray Leaf Bugs.) ....*Piesmidae*

Rostrum three segmented. Membrane without trace of veins. Upper surface often distinctly punctate but not areolate. (Royal Palm Bugs.) (*Thaumastocoridae*) ....*Thaumastotheriidae*

30. Rostrum three segmented. Hind legs enlarged for jumping... 31

Rostrum four segmented. Hind legs not as above... 32

31. First two antennal segments very stout, the apical two slender. Ocelli widely separated, nearer to eyes than to each other. Very small species, less than 2 mm, long. (Jumping Ground Bugs.) .... *Cryptostemmatidae*

Antennal segments all about equally thick. Ocelli approximate, nearly contiguous. Larger than 2 mm. (Shore Bugs.) ... . *Saldidae*

32. Body and appendages long and slender. First antennal segment very long, bent just beyond its bae..*Neididae*

Body and antennae not as above.... 33

33. Membrane with numerous more or less distinct longitudinal veins. Antennae often inserted high on the head, above a line drawn from middle of eyes to front end of bucculae (including CORIZIDAE). (Squash Bugs.).... .... *Coreidae*

Membrane with only four or five longitudinal veins, the inner two sometimes joined by a cross vein. Antennae usually inserted either on or below a line drawn from middle of eyes to front end of bucculae. (Chinch Bugs.).. *Lygaeidae*

34. Scutellum very large, U-Shaped, longer than the corium, covering a large part of the abdomen.. 35

Scutellum smaller, triangular or moderately rounded at apex, shorter than the corium.. 37

35. Tibiae with several rows of stout spines. Body small, subrounded, and shining black, often marked with yellow or red. (Negro Bugs.) *(Corimelaenidae)*.. *Thyreocoridae*

Tibiae not strongly spinose. Color and body form not as above 36

36. Sides of pronotum with a prominent tooth or lobe in front of the humeral angles and another at anterior angles.........................*Podopidae*

Sides of pronotum without such protuberances. (Shield Bugs.) ....*Scutelleridae*

37. Tibiae armed with several rows of stout spines. Front legs fossorial. (Ground Bugs, Burrower Bugs.).... .... *Cydnidae*

Tibiae unarmed or with only a few small or scattered spines. Front legs not fossorial. (Stink Bugs, Solier Bugs.) ....*Pentatomidae*

Order - ORTHOPTERA

Key to sub-orders

1. Antennae shorter than body (<30 segments) auditory organs when present, near the base of abdomen, ovipositor short, tarsi usually triarticulate (Locust and Grasshoppers) .... *Caelifera*

Antennae longer than body (>30 segments) auditory organs when present, on fore tibia, ovipositor long and needle shaped or sword shaped, tarsi usually with 4 segments. Long horned Grasshoppers Katydids, Crickets .... *Ensifera*

Sub-order - CAELIFERA

1. Pronotum greatly extended, often beyond the tip of the abdomen fore or middle tarsi 2 segmented, hind tarsi 3 segmented arolia absent, tegmina very small or scale like on side of thorax, peculiar small insects .... *Tettigidae*

Pronotum not extended beyond the tip of abdomen, tegmina not-vestigial, arolia present, antennae not longer than fore femora 2

2. Body stick like, wings either absent or vestigial, antennae with 8 segments, head conical .... *Proscopidae*

Not as described above.. 3

3. Hind femur not strongly thickened, abdomen inflated, pronotum very large .... *Pneumoridae*

Hind femur greatly swollen at base and lengthened, abdomen not noticeably swollen grass-hoppers, locusts.. *Locustidae*

Sub-order - ENSIFERA (Long horned grasshoppers)

1. Tarsi 4 segmented, body cylindrical or compressed and robust, ovipositor when present, sword shaped or sickle shaped 2

Tarsi 3 segmented or less, body somewhat depressed, ovipositor when present, spear shaped or awl shaped 3

2. Ovipositor sword shaped or sickle shaped, body compressed auditory organ often present on fore tibia, winged or without wings .... *Tetigonidea (Katydids)*

Ovipositor inconspicuous, body robust, auditory organ not present on foretibia, without wings or partially developed wings (Sand crickets, camel crickets).. *Stenopelmatidae*

3. Fore tibia normal, usually with auditory organ on each side, ovipositor well developed, spear shaped or awl shaped (Crickets, Ant crickets, Tree crickets) .... *Gryllidae*

Fore tibia greatly enlarged or fossorial, ovipositor not exposed..................... 4

4. Antennae filiform, many segmented, 2 ocelli present tarsi 3 segmented hind femur not enlarged for jumping, burrow in the soil, large insects body clothed with short bristles (Mole crickets).................... *Gryllotalpidae*

Antennae short, 12 segmented or less, 3 ocelli present, Fore and middle tarsi 2 segmented, hind tarsi 1 segmented or obsolate, small insects will small bodies, hind femur greatly enlarged for jumping (Pigmy molecrickets)... *Tridactylidae*

Antennae long tegnina long and rolled into a spiral at the tip tarsi have curious expansions (Schizodactylus) .... *Schizodactylidae*

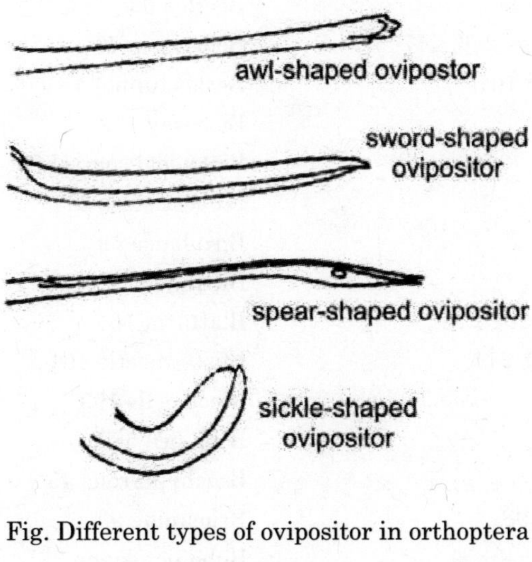

Fig. Different types of ovipositor in orthoptera

SUBJECT INDEX